Praise for *Wind Energy Basics*

"This second edition of Gipe's *Wind Energy Basics* is a much-needed critique of the state of small wind today. Gipe advances what will inevitably be a growth industry in the U.S.: community wind projects, based on the successful models in many parts of Europe over the past three decades. Paul also advocates for equitable feed-in tariffs for wind to level the playing field for wind turbines that work. Finally, Gipe wades through the numerous Internet wonders and surrounding hype that are doing more harm than help for prospective turbine owners. *Wind Energy Basics* is a must read, and reread."

—MICK SAGRILLO, "Advice from an Expert" columnist for the American Wind
Energy Association and coauthor of *Power from the Wind*

"Without abandoning the needs of individuals aiming for energy independence, Gipe wisely promotes community-scale wind power in his new book. He is not only an unrivalled expert, but an excellent teacher as well."

—PETER BARNES, author of *Climate Solutions*

"If you want straight talk on wind electricity, with no bull, seek out Paul Gipe. Not beholden to any company or segment of the industry, Paul tells it like it is. His no-nonsense book will steer you in the right direction—away from fantasy and failure and toward a successful wind-electric system."

—IAN WOOFENDEN, Senior Editor, *Home Power* magazine

"Gipe's call for an ethical energy policy in *Wind Energy Basics* is a message that North American politicians should heed. The people deserve nothing less."

—GLEN ESTILL, past President of the Canadian Wind
Energy Association and successful wind entrepreneur

"Paul Gipe, the country's leading expert on small-scale and locally owned wind energy, has written the how-to manual for those who want to literally bring power to the people."

—DAVID MORRIS, Vice President, Institute for Local Self-Reliance, and
author of *Seeing the Light: Regaining Control of Our Electricity System*

"Paul Gipe is an independent, opinionated voice on wind energy, cutting right to the core on almost any wind energy topic. He analyzes the issues with uncompromising standards. With Paul's journalistic background and years in the worldwide wind industry, he has no trouble sharing the story as he sees it, encouraging all to explore business models and policies that offer something more, something for all of us. Paul is a visionary on the energy front, presenting a compelling case for change."

—Lisa Daniels, Executive Director, Windustry

"Gipe provides the reader with an informative and easy-to-understand guide to small and micro wind systems for the generation of energy.... *Wind Energy Basics* is a 'must' for environmentally supportive advocates seeking to establish nonpolluting energy resources for themselves, their families, and their businesses."

—*Midwest Book Review* (refers to the first edition of *Wind Energy Basics*)

"[A] wonderful primer for all but the professional wind enthusiast."

—Trevor Robotham, proprietor of Sun Wind & Power (refers to the first edition of *Wind Energy Basics*)

WIND
ENERGY
BASICS

WIND ENERGY BASICS

A Guide to Home- and Community-Scale Wind Energy Systems

SECOND EDITION

PAUL GIPE

Chelsea Green Publishing Company
White River Junction, Vermont

Project Manager: Emily Foote
Developmental Editor: Matthew Scanlon
Copy Editor: Laura Jorstad
Proofreader: Helen Walden
Indexer: Lee Lawton
Designer: Peter Holm, Sterling Hill Productions

Printed in the United States of America
First printing, March, 2009
10 9 8 7 6 5 4 3 2 1   09 10 11 12 13

Our Commitment to Green Publishing

Chelsea Green sees publishing as a tool for cultural change and ecological stewardship. We strive to align our book manufacturing practices with our editorial mission and to reduce the impact of our business enterprise on the environment. We print our books and catalogs on chlorine-free recycled paper, using vegetable-based inks whenever possible. This book may cost slightly more because we use recycled paper, and we hope you'll agree that it's worth it. Chelsea Green is a member of the Green Press Initiative (www.greenpressinitiative.org), a nonprofit coalition of publishers, manufacturers, and authors working to protect the world's endangered forests and conserve natural resources.

Wind Energy Basics, Second Edition, was printed on Renew Matte, a 20-percent postconsumer recycled paper supplied by RR Donnelley.

Library of Congress Cataloging-in-Publication Data
Gipe, Paul.
   Wind energy basics : a guide to distributed wind energy / Paul Gipe. -- 2nd ed.
      p. cm.
   Includes bibliographical references and index.
   ISBN 978-1-60358-030-4 (alk. paper)
   1. Wind power. 2. Distributed generation of electric power. I. Title.

   TJ820.G55 2009
   621.31'2136--dc22

                           2008053185

Chelsea Green Publishing Company
Post Office Box 428
White River Junction, VT 05001
(802) 295-6300
www.chelseagreen.com

Disclaimer

The installation and operation of wind turbines entail a degree of risk. Always consult the manufacturer and applicable building and safety codes before installing or operating your wind power system. For wind turbines that will be interconnected with the electric utility, always check with your local utility first. When in doubt, ask for advice. Suggestions in this book are not a substitute for the instructions of wind turbine manufacturers, or regulatory agencies, or for common sense. The author assumes no liability for personal injury, property damage, or loss arising from information contained in this book.

For Ed Wulf, a man who saw the future.

CONTENTS

LIST OF TABLES

Preface

Wind Energy Basics: A Guide to Home- and Community-Scale Wind Energy Systems is a revision of my 1999 *Wind Energy Basics: A Guide to Small and Micro Wind Systems*. As the new title suggests, this version expands the scope of the earlier book to include commercial-scale wind turbines used in distributed applications. As such, this book includes wind turbines of all sizes. This edition makes a distinction between large numbers of commercial-scale wind turbines used in central-station power plants, or wind farms, and wind turbines used singly or in small clusters both on and off the grid.

This book is not by any means exhaustive, nor is it intended to be. In the more than three decades I've worked with wind energy, the field has grown so vast that it's no longer possible to confine the technology within the covers of one book, even after limiting it to distributed applications.

Wind Energy Basics is intended as a companion to *Wind Power: Renewable Energy for Home, Farm, and Business* (Chelsea Green, 2004). In 1999 *Wind Energy Basics* introduced micro and mini wind turbines and explained how to install and use them. This version introduces the concept of community wind, in which groups of people invest in large wind turbines that produce commercial quantities of electricity for sale to the grid. While a seemingly novel concept in North America, it is quite common

in Denmark, Germany, and increasingly France. In community wind, farmers, small businesses, and groups of community-minded citizens band together to develop—for profit— "their" wind resources. It's as if they're saying, *Renewable energy is far too important to be left to the electric utilities alone. We have a responsibility for our own future. We can and will develop our own wind resources for our own benefit and for the benefit of our communities.* By proving that it can be done, Germans and Danes have served as models for us in North America.

All books, even small ones, require the help and cooperation of many people. I am thankful to the many wind turbine manufacturers worldwide who answered my frequent queries about their products, and to Mick Sagrillo, Hugh Piggott, Ian Woofenden, and Ken Starcher for their comments and insights on small wind turbine design.

And I am truly grateful to the Folkecenter for Renewable Energy and the people of Denmark for a fellowship to study the distributed use of wind energy in northwest Jutland.

The people of Ontario and the members of the Ontario Sustainable Energy Association also deserve a special note of thanks for having the faith in themselves and their communities to support the development of the most progressive renewable energy policy in North America in more than two decades. Since Ontario launched its groundbreaking Standard Offer Contract program, a policy modeled after those in Denmark and Germany, the revolutionary idea has caught on throughout North America.

Soon, I hope, we'll see communities across the continent clamoring for the right to connect their wind turbines to the grid—and their solar panels and biogas plants as well—and be paid a fair price for their electricity. Only then will we see the promise of renewable energy fulfilled.

God vind! (Good wind!)

PAUL GIPE
Tehachapi, California
August 2008

Overview

Galloping climate change and dwindling supplies of fossil fuels are driving ever-greater interest in renewable energy. As a result, wind energy is booming worldwide. Not since the heyday of the American farm windmill has wind energy grown at such a dramatic pace. Today there are more than 100,000 commercial-scale wind turbines and untold thousands of small wind turbines spinning out more than 160 terawatt-hours (billion kilowatt-hours) of electricity annually, from the steppes of Mongolia to the shores of the North Sea.

Relative Size

In wind energy, size—especially rotor diameter—matters. More specifically, the area swept by the wind turbine's rotor—or the area of the wind it intercepts—is the single most important aspect of a wind turbine.

Wind turbines range in size from Southwest Windpower's 200-watt Air Breeze, a micro turbine, which uses a rotor only 1.2 meters (2.8 ft) in diameter, to Enercon's 6,000 kW giant, with a rotor spanning 126 meters (400 ft). There is no ironclad rule on what constitutes a small or large wind turbine. Size designations are somewhat arbitrary. Clearly the Air Breeze is small, and Enercon's 6-megawatt turbine is not.

Wind turbines of any size can be used in distributed applications either singly or in small groups.

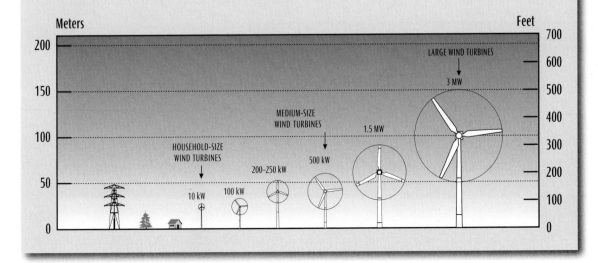

While attention in North America until now has been on the giant wind farms springing up across the breadth of the continent like giant mushrooms, there's another, often overlooked side to wind energy: distributed generation, or putting wind power where people live and work.

Distributed wind generation is booming, too, mostly in Europe, where the use of wind turbines in or near cities and villages is commonplace. But a growing number of small wind turbines are also finding homes on sailboats, at remote cabins, or at new homesteads at the end of the utility's lines or even off the grid entirely.

The focus of the original edition of *Wind Energy Basics* was small wind turbines. This version broadens its scope to include large wind turbines used in distributed applications. Consequently, this edition of *Wind Energy Basics* expands the number of wind turbine size classes over those used in the earlier edition.

Wind Turbine Size Classes

Some wind turbines are so small you can pick them up in your hands. Mongolian nomads carry these micro turbines on horseback from one encampment to the next. Other wind turbines are so large you can see them from commercial airliners as you streak across the sky (see table 0-1, Wind turbine size classes).

Though we often use the power rating of a wind turbine as shorthand for its size, this can be very misleading. Wind turbine size classes depend primarily upon the diameter of the rotor, or, more correctly, the area swept by the rotor. And this is true regardless of orientation, whether we are describing conventional wind turbines or wind turbines that spin about a vertical axis.

While their contributions may be small in absolute terms, small wind turbines make a big difference in the daily lives of people in remote areas around the globe. Small wind turbines may produce only a few tens of kilowatt-hours of electricity per month, but this electricity goes much farther and provides as much, if not more, value to those who depend upon it as the generation of large wind turbines in areas served by utility power.

Typically, small wind turbines encompass machines producing anywhere from a few watts

Table 0-1: Wind Turbine Size Classes

	Rotor Diameter		Swept Area		Standard Power Rating*	
	m	~ft	m²	~ft²	kW	kW
Micro	0.5 – 1.25	2 – 4	0.2 – 1.2	2 – 13	0.04	0.25
Mini	1.25 – 3	4 – 10	1.2 – 7.1	13 – 76	0.25	1.4
Household	3 – 10	10 – 33	7 – 79	76 – 845	1.4	16
					Typical Manufacturer Power Rating	
Small Commercial	10 – 20	33 – 66	79 – 314	840 – 3,400	25	100
Medium Commercial	20 – 50	66 – 164	314 – 1,963	3,400 – 21,100	100	1,000
Large Commercial	50 – 100	164 – 328	1,963 – 7,854	21,100 – 84,500	1,000	3,000

*Std. Power Rating for micro, mini, and household-size wind turbines = swept area x 200 W/m².

Units of Measurement

In the United States we still rely on the old English or "Imperial" system of measurements. Canadians have successfully made the transition to metric. Continental Europeans use metric measurements exclusively. The rotor diameter of wind turbines, because they are sold internationally, is nearly always given in meters. (Some American manufacturers of small wind turbines give rotor diameter in both feet and meters.) Most of those who work with wind energy are accustomed to using the metric system, especially when referring to the size of a wind turbine by its rotor diameter.

When the size of a wind turbine is mentioned in *Wind Energy Basics,* the measurement will be given in meters. If you have an aversion to metric units, don't panic. The approximate conversion to feet will accompany in parentheses.

To calculate the power in the wind and to estimate the amount of energy a wind turbine is likely to produce, it will make your life a lot easier to use the metric system. If wind speed is given in mph (miles per hour) or in knots, simply convert the speed to the metric system's m/s (meters per second).

Here are some useful conversions.

$$1 \text{ m/s} = 2.24 \text{ mph}$$
$$1 \text{ m/s} = 1.94 \text{ knots}$$
$$1 \text{ meter} = 3.28 \text{ feet}$$

If you have a hard time visualizing the rotor diameter of wind turbines in meters, here are some simple approximations.

1 meter ~ 3 feet (the size of the Marlec's Rutland 910)

2.5 meters ~ 8 feet (about the size of Bergey's XL.1)

15 meters ~ 50 feet (the size of an old Vestas V15)

30 meters ~ 100 feet

50 meters ~ 150 feet (about the size of WindShare's Lagerwey turbine in Toronto)

100 meters ~ 300 feet (among the biggest wind turbines made)

---

to 10–20 kW. Rotors reaching 10 meters (30 ft) in diameter drive wind turbines at the upper end of this range. Small wind turbines can be subdivided further into micro wind turbines—the smallest of small turbines—mini wind turbines, and household-size wind turbines.

In *Wind Energy Basics* we classify micro turbines as those from 0.5 to 1.25 meters (2–4 ft) in diameter. These machines include Southwest Windpower's 200-watt Air Breeze as

well as Ampair's model 300. Both use rotors 1.2 meters in diameter and intercept about 1 square meter of the wind stream (see figure 0-1, Micro wind turbine).

Mini wind turbines are slightly larger and span the range between the micro turbines and the bigger household-size machines. They vary in diameter from 1.25 to 3 meters (4–10 ft). Popular turbines in this category include Southwest Windpower's Whisper 100 as well

Figure 0-1. Micro Wind Turbine. The Ampair 300, a new micro turbine from venerable Ampair, uses a rotor 1.2 meters (4 ft) in diameter.

as Bergey Windpower's XL.1. The Whisper 100 uses a 2.1-meter (7 ft) rotor and intercepts 3.6 square meters (m²) while the bigger XL.1 uses a 2.5-meter (8 ft) rotor that sweeps 4.9 m². Thus, these two turbines are four to five times bigger than the Air Breeze or Ampair 300 (see figure 0-2, Mini wind turbine).

Household-size wind turbines (a translation of the Danish term *hustandmølle*) are the largest of the small wind turbine family. As you would expect, wind turbines in this class span a wide spectrum. They include models as small as Southwest Windpower's Skystream with a

Figure 0-2. Mini Wind Turbine. The Whisper H40 (now designated the Whisper 100) operating at the Wulf Test Field in California's Tehachapi Pass.

Figure 0-3. Household-Size Wind Turbine. The Skystream 3.7 uses a rotor 3.7 meters (12 ft) in diameter and intercepts about 10 m² of the wind stream. (Detronics, www. detronics.net)

rotor 3.7 meters (12 ft) in diameter, as well as the Bergey Excel that uses a rotor 7 meters (23 ft) in diameter and weighs in at nearly 500 kilograms (1,000 lbs). The Skystream sweeps 10 m², whereas the Bergey sweeps nearly 4 times more area than the Skystream and 40 times more than the Air Breeze (see figure 0-3, Household-size wind turbine).

Small commercial turbines, such as Entegrity's EW50, are intended for farms, schools, and small businesses. They range in diameter from 10 to 20 meters (30–70 ft) and sweep up to 300 m². Entegrity's 50 kW turbine, patterned after hundreds of similar turbines used in California during the 1980s, intercepts 175 m². Turbines in this class are capable of producing from 50 kW to 100 kW (see figure 0-4, Small commercial-scale wind turbine).

Figure 0-4. Small Commercial-Scale Wind Turbine. The Windmatic 15S at Dutch Valley Produce, a Hutterite colony near Pincher Creek, Alberta. The used 65 kW turbines were transplanted from California.

Figure 0-5. Medium-Size, Commercial-Scale Wind Turbine. A Vestas V27 in a city park overlooking New Zealand's capital, Wellington. This wind turbine has been among the world's most productive for more than a decade.

Figure 0-6. Large Wind Turbine. Sky Generation's Vestas V80 on Ontario's Bruce Peninsula. Though the 1.8 MW turbine is commercial in scale, it is connected at the distribution voltage. This turbine is one of a cluster of three machines all similar in size and nicely illustrates the way large wind turbines can be used effectively in distributed generation.

Medium-size wind turbines are those used for commercial applications such as farms, factories, businesses, and small wind farms. They can range from 20 to 50 meters (70–160 ft) in diameter and sweep as much as 3,000 m². Turbines in this class can be rated from 100 kW to more than 1,000 kW (see figure 0-5, Medium-size, commercial-scale wind turbine).

Large commercial-scale turbines are the machines found in modern wind power plants. Though huge on a human scale, they can be found singly or in small clusters in or near cities

Using www.wind-works.org

This version of *Wind Energy Basics* is designed to use with access to the World Wide Web, especially the author's Web site: www.wind-works.org. Here's a quick guide to what you'll find there:

• **Books:** A list of Paul Gipe's other books on wind energy and related topics, including *Wind Power: Renewable Energy for Home, Farm, and Business*. Any corrections, changes, updates, or addendums to this book can be found here.

• **Large wind:** Articles and commentary for wind professionals and renewable energy advocates on commercial wind development. This section contains articles and links on wind statistics, economic calculations, accidents, and book reviews.

• **Small wind:** Articles, commentary, and links on small and household-size wind turbines, including small turbine testing, inventions, questionable turbines, economic calculations, safety, and book reviews. A few of the articles on small wind turbines are in French and German.

• **Co-op wind:** Articles and commentary on community wind; wind energy and the environment; landowner easements; and royalty payments.

• **Feed laws:** One of the world's most extensive collections of articles, papers, and presentations on electricity feed-in tariffs, Advanced Renewable Tariffs, and Renewable Energy Payments. This section includes tables of renewable energy tariffs worldwide, and a country-by-country discussion of policy developments.

• **Solar energy:** Articles and commentary on the development of solar photovoltaics and solar thermal systems.

• **Wulf Field:** Description of the Wulf Test Field in the Tehachapi Pass, the turbines tested, and measurements from those tests.

• **Workshops:** Description of seminars on wind energy and Advanced Renewable Tariffs by Paul Gipe and a schedule of upcoming events.

• **Links:** A list of recommended associations, organizations, consultants, and individuals working with wind energy.

---

throughout Europe. For a comparison of the scale, consider the Vestas V80, a wind turbine variously rated from 1,800 kW to 2,000 kW, depending upon where it is used in the world. This wind turbine sweeps more than 5,000 m²; that is, the V80 is 5,000 times bigger than the Air Breeze or the Ampair 300 (see figure 0-6, Large wind turbine).

What's New

This version of *Wind Energy Basics* has been extensively updated to include topics of increasing interest to North American consumers and wind energy advocates alike.

• Urban wind. Does it make sense?

Power and Energy—Knowing the Difference

In casual conversation, we use the terms *power* and *energy* interchangeably. But knowing the difference between the two can save you a lot of headaches—and a lot of money.

Energy is the ability to do work or the amount of work actually performed. For our purposes here, energy is given in kilowatt-hours (kWh) of electricity produced by a wind turbine or consumed in a home or business. When most people pay their utility bill, they pay for the electricity they consumed in kWh.

Power is the rate at which energy is generated or consumed, that is, kilowatt-hours per hour (kWh/h) or kilowatts (kW). One kilowatt is 1,000 watts (W). One megawatt is 1,000 kilowatts or 1 million watts.

The distinction between kilowatts and kilowatt-hours is critically important. Knowing the difference can keep you from being confused by a wind turbine's size in kilowatts (or for very small wind turbines, watts), and how much energy, in kilowatt-hours, it will actually produce. Some unscrupulous manufacturers play upon the public's ignorance of this distinction and give their wind turbines a very high "power rating" when the actual turbine is unlikely to deliver as much electricity as a competitor with a low power rating.

The "power" rating of a wind turbine is an unreliable and often very misleading shorthand for how much energy a wind turbine will capture.

In *Wind Energy Basics* the emphasis is on "energy." The most reliable indicator of how much electricity a wind turbine will generate is its rotor diameter.

• Building integrated wind. Is it real or not?
• Rooftop mounting. Should you avoid it?
• New vertical-axis wind turbines. Are they ready?
• Fantasy wind turbines. How to spot them.
• Ducted turbines. Can they deliver?
• Community wind. A not-so-new way to harness the wind.
• Feed-in tariffs. Can they power a renewables revolution?

The most significant change is the addition of a new chapter on community wind, and why this can be an exciting option for many who might otherwise struggle to put a small wind turbine in their backyards—or, worse, on their roofs. Another departure from the earlier version is a concluding chapter on a policy option that can make all this possible: Advanced Renewable Tariffs and the feed-in tariffs that make them work.

What's Remained the Same

Wind energy—especially when it comes to small wind turbines—has been plagued with hustlers and fast talkers "selling wind on hope and hype," as one wag put it. The situation has only gotten worse as the use of the Internet has grown. It's never been easier to pawn off fantasy wind turbines—or as Mick Sagrillo, the wind sage of Wisconsin, likes to call them, "Internet wonders"—onto an unsuspecting public. As in the previous version, this edition of *Wind Energy Basics* frankly tells you what works, what doesn't, and what to avoid.

However, this remains a slim book on the basics. For more detail, see the companion volume *Wind Power: Renewable Energy for Home, Farm, and Business*, 500 pages of all the number-crunching most people interested in wind energy will ever need. Professionals can find reviews of engineering textbooks on wind energy at www.wind-works.org.[1]

In the next chapter we'll explore wind technology, then move on to the basics of wind energy and methods for estimating how much electricity a wind turbine may produce.

Wind Power: Renewable Energy for Home, Farm, and Business

Chelsea Green's companion book by Paul Gipe is an extensive 500-page treatment of the topics touched on in *Wind Energy Basics*. Profusely illustrated and with numerous tables, *Wind Power* has become the reference work for the amateur as well as the budding professional interested in the technology and the promise of wind energy.

---

1. www.wind-works.org/articles/large_turbines.html#Reviews.

Technology

Wind Turbine Technology

Wind turbines often confuse the uninitiated. They come in a bewildering variety of shapes and sizes. At one time the situation was even worse: Practically every conceivable form was on the market. In the past two decades, though, the technology has steadily evolved toward a common configuration: rotors of three slender blades spinning about a horizontal axis upwind of the tower. Though there are important variants, such as new vertical-axis turbines, the differences among most wind turbines are subtle. These include differences in how they generate electricity and how they are controlled in high winds. As Real Goods' Doug Pratt says, "Wind [energy] isn't something that's mysterious anymore."

That's not to say innovation is dead. The technology is advancing for both small and large wind turbines at a steady, incremental—some might say plodding—pace. We've found that's what works best. Despite all the "revolutionary" new wind technology that's been promised in the past 30 years, there have been no breakthroughs, no paradigm-shifting developments, no game-changing marvels that have forced us to rethink how to build wind turbines that work reliably for decades on end.

Of course, inventors have tried. With the price of oil creeping ever higher and natural gas following suit, wind turbine inventions are flooding the Web. It seems that every week a "new" wind turbine appears. Most are "Internet wonders," as Mick Sagrillo derisively calls them: digital apparitions that cause a flurry of media interest, then fade from view until the next mirage appears on the horizon.

Real wind turbines are made of nuts and bolts, steel and wood, fiberglass and concrete. You can touch them, feel them—hear them, too. This chapter examines real wind turbines, the ones you can see and hear, but it also offers tips on how to spot—and avoid—the fantasy machines.

Wind turbines have been built in nearly every conceivable configuration. Today nearly all wind turbines are upwind, horizontal-axis turbines whose rotor spins in front of the tower, about a line parallel with the horizon. But it hasn't always been so. Even today, three decades after the modern wind energy renaissance, new vertical-axis wind turbines continue to appear.

Rotor Orientation

The rotor of a wind turbine can spin about a horizontal or vertical axis—thus the designations HAWT, for horizontal-axis wind turbine, and VAWT for vertical-axis wind turbine. Today we consider horizontal-axis wind turbines to be conventional technology. VAWTs remain rare, and because so few are in use, the public is easily captivated by what

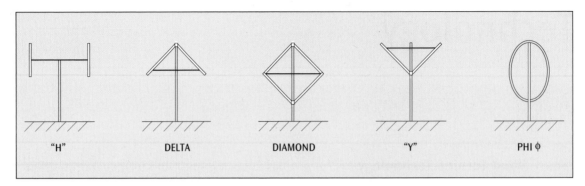

"H"   DELTA   DIAMOND   "Y"   PHI ϕ

Figure 1-1. Vertical-Axis Wind Turbines. There are several VAWT configurations, and all have been tried at one time or another.

seems a "new" technology. However, the vertical-axis orientation probably predates conventional wind turbines. Persian panemones were in use in what is now Afghanistan long before the European or Dutch windmill began sprouting in England and the Low Countries.

French aeronautical engineer Georges Jean-Marie Darrieus developed what we consider the "modern" vertical-axis wind turbine in the 1920s. His inventions include the H-rotor, the diamond rotor, and what has become the classic Darrieus or "eggbeater" shape, the phi-configuration (Φ) (see figure 1-1, VAWT configurations). The Darrieus design languished until the 1970s, when it was rediscovered by Canadian researchers. Since then the design has fallen in and out of favor in Canada and elsewhere.

The most successful VAWT is the widely used cup anemometer. It's seen almost everywhere wind energy is being used or measured. Because of its simplicity, just three cups mounted about a vertical axis, it is common for backyard inventors to rediscover it and announce some new "breakthrough." Unfortunately, the cup anemometer's simplicity is its principal drawback. The cups are nothing more than wind buckets. They are not airfoils or wings.

The cups use aerodynamic drag to drive the materially intensive rotor. Wind turbines derived from cup anemometers, and other VAWTs like them, can be used in small, battery-charging applications or for pumping water, but will be far less cost-effective than conventional wind turbines (see figure 1-2, Cup anemometer).

Another simple vertical-axis wind turbine that has found limited application for battery charging at remote locations, such as marker buoys, is the Savonius or S-rotor. Like the cup anemometer, Savonius rotors are useful where their simplicity and ruggedness are required. This design was common in back-to-the-land magazines during the 1970s as a wind turbine that could be easily built at home or in a Third World shop (see figure 1-3, Savonius VAWT). There's more on new VAWTs later in this chapter.

What we consider conventional wind turbines first appeared along the shores of the Mediterranean, then eventually along the North Sea and the English Channel. Dutch or European windmills reached a surprisingly high degree of sophistication by the 18th and 19th centuries. Like most modern horizontal-

Figure 1-2. Cup Anemometer. The cup anemometer uses aerodynamic drag on the "cups" to propel its vertical-axis rotor. It's far less efficient than conventional wind turbines like the one in the background.

Figure 1-3. Savonius VAWT. Finnish manufacturer Windside builds small Savonius rotors for remote, battery-charging applications.

axis wind turbines, European windmills placed the rotor in front of the tower, thus requiring a means to point the rotor into the wind (see figure 1-4, Horizontal-axis wind turbines).

Early windmills required the miller to reorient the rotor manually as the wind changed directions. Later versions used mechanical devices, fantails, to do the job automatically. Fantails were the forerunners of the electrical yaw drives used today to keep large wind turbines facing the wind.

Following the example of the American water-pumping windmill widely used in the late 19th century, most small wind turbines today use a tail vane to passively orient the rotor into the wind.

In the 1970s and 1980s many small wind turbines were designed to passively orient themselves downwind of the tower. This seemed modern at the time, even though farm windmills had used the same concept in the mid-19th century. At one point there were literally thousands of such turbines in use.

Today only one small wind turbine, Southwest Windpower's Skystream 3.7, operates in this manner. Similarly, only one small

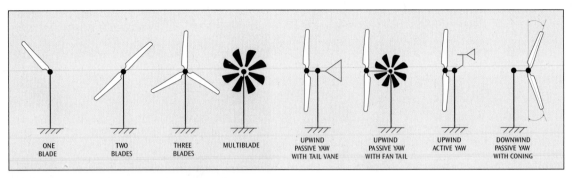

Figure 1-4. Horizontal-Axis Wind Turbines.

Figure 1-5. Downwind Wind Turbine Rotor. The rotor on Southwest Windpower's Skystream 3.7 is oriented downwind of the tower. The force of the wind keeps the rotor there. (Appalachian State University, www.wind.appstate.edu)

commercial turbine, the Entegrity 15, passively orients itself downwind of the tower (see figure 1-5, Downwind wind turbine).

Downwind HAWTs have an unusual propensity to "walk" around the tower under certain wind conditions. This can be troublesome in high winds, if the blades are so flexible that they can bend back far enough to hit the tower.

One, Two, or Three Blades

Whether it's a VAWT or a HAWT, a wind turbine needs only one blade to convert the energy in the wind stream. Most Darrieus turbines of the 1980s used two blades, though Alcoa developed a three-blade model that was never commercialized. Many of the new VAWTs on the market use three blades.

Most conventional wind turbines use two or three blades, though there have been occasional one-blade and four-blade models briefly on the market (see figure 1-6, One-blade HAWT).

Designers of conventional wind turbines have had a long and bitter debate about the merits of using two or three rotor blades. The only advantage of two blades over three is that two are cheaper. But it's a case of penny wise and pound foolish, as the British would say. Turbines with three blades run more smoothly than two, and that usually means they will last longer. After long experience with wind turbines, the US Department of Agriculture's Nolan Clark opts for three-blade rotors over the two-blade machines still occasionally found.

Among small wind turbines, only Southwest Windpower's Whisper 500 (formerly the Whisper 175) still uses a two-blade rotor. The household-size wind turbine uses a 4.6-meter (15 ft) diameter rotor. Only a few manufacturers of large wind turbines continue to develop two-blade rotors. The most successful is Geoff Henderson's Windflow 500, a 33-meter (110 ft) diameter wind turbine being installed in New Zealand.

Blade Materials

Wind turbine blades can be made out of wood, cloth, steel, aluminum, and fiberglass—and have been, at one time or another. Few manufacturers today use metal blades. Aluminum is an especially poor choice because of its propensity to metal fatigue. Nearly all use composites of fiberglass, carbon fiber, or wood.

Robustness

Wind turbines work in a rugged environment. You quickly appreciate this when you watch a small wind turbine struggling through a gale. There's no foolproof way to evaluate the robustness of wind turbine designs. You certainly can't rely on the manufacturer's pronouncements. No manufacturer is going to tell you that its turbine is only suitable for light winds.

In general, heavier small wind turbines have proven more rugged and dependable than lightweight machines. Wisconsin's Mick Sagrillo is a proponent of what he calls the "heavy-metal

Figure 1-6. One-Bladed HAWT. Riva Calzoni 33 m (108 ft) diameter, 300 kW, one-blade wind turbine in the central Apennines of Italy in the late 1990s. There are no one-blade wind turbines on the market today.

school" of small wind turbine design. Heavier, more massive turbines, he says, typically run longer. Heaviness, in this sense, is the weight or mass of the turbine relative to the area swept by the rotor. By this criterion, a turbine that has a relative mass of 10 kg/m² may be more robust than one with a specific mass of 5 kg/m². While not a surefire way of determining whether one wind turbine will last longer than another, it remains a useful tool (see table 1-1, Specific Tower Head Mass for Selected Small Turbines).

For example, over the years we've found that

Table 1-1: Specific Tower Head Mass for Selected Small Turbines						
Manufacturer	Model	Rotor Diameter m	Area m²	Std. Power Rating* kW	Tower Top Mass kg	Specific Mass kg/m²
Proven Wind Turbines	2.5	3.5	9.6	1.9	190	20
Abundant Renewable Energy	110	3.6	10.2	2.0	143	14
Bergey Windpower	Excel-S	7.0	38.5	7.7	463	12
Southwest Windpower	Skystream	3.7	10.9	2.2	77	7
Southwest Windpower	Whisper 100	2.1	3.6	3.6	21	6
Southwest Windpower	Whisper 500	4.6	16.4	3.3	70	4

*Standard Power Rating: 200 W/m².

the Bergey Excel (12 kg/m²) will work reliably for long periods. Southwest Windpower's Whisper 500 (4 kg/m²), on the other hand, has been far more trouble-prone. Similarly, Proven's turbines (20 kg/m²) have been used extensively in windy Scotland and on the Falkland Islands. (There will be more discussion of longevity of small turbines in chapter 8, Investing in Wind Energy.)

While there are no standards for small wind turbines that will tell you whether one wind turbine is better designed for low-wind sites than for windy sites, there are such standards for commercial-scale wind turbines. Large wind turbines are designed for specific wind classes (see table 1-2, Large Wind Turbine Classes). Wind Class IV turbines, for example, use large rotors relative to the same wind turbines intended for windy sites, such as Class I.

Using a Class IV wind turbine at a Class I site will likely lead to early failure, as the rotor is too large for the wind turbine and the wind conditions—it will overpower the drive train and could ultimately lead to a catastrophic failure of the rotor. That's why you would never use a Class IV turbine at a Class I site like New Zealand's

Tararua Ranges northeast of Wellington—one of the windiest sites in the world.

It is for this reason that some refer to turbines designed for Class IV conditions as low-wind-speed turbines. Class IV conditions are typical of much of the American Midwest, though there are windier locales, such as Minnesota's Buffalo Ridge, where low-wind-speed turbines are unsuitable.

Overspeed Control

All turbines must have some means of controlling the rotor in high winds. This is fundamental to wind turbine design. Gale-force winds are highly destructive, and a wind turbine must be designed to operate safely under such conditions or else in some manner turn itself either off or out of the wind. If not, there is the danger that the rotor will go into "overspeed," which could eventually destroy the wind turbine. Overspeed control is one of the characteristics that sets different wind turbines apart.

Danes go further. They learned 30 years ago that all wind turbines must have an "aerodynamic" means of limiting rotor power in an emergency. The Danish wind turbine owners'

Table 1-2: Large Wind Turbine Classes								
	I		II		III		IV	
	m/s	mph	m/s	mph	m/s	mph	m/s	mph
Annual Average Wind Speed	10	22	8.5	19	7.5	17	6	13
Reference Wind Speed	50	112	42.5	95	37.5	84	30	67
50-year Gust Speed	70	157	59.5	133	52.5	118	42	94

Notes: 10-minute averages, hub height wind speed.

association was formed in part to force Danish manufacturers to abide by this requirement. How did they do this? They simply refused to buy any wind turbine that didn't have it. They did so because they had already lost too many wind turbines to winter gales. Some consider their action to be one of the reasons the Danish wind industry grew to dominate the world wind turbine market for more than two decades.

will not destroy itself. But without aerodynamic overspeed controls, the likelihood increases that a wind turbine will not survive an emergency intact. This lesson has been continually repeated over the decades of modern wind power.

Examples of aerodynamic overspeed controls include furling the rotor toward the tail vane in small wind turbines, the centrifugally activated pitchable blade tips on early Danish wind

---

Danes go farther. They learned 30 years ago that all wind turbines must have an "aerodynamic" means of limiting rotor power in an emergency.

Danish consumers literally saved Danish wind turbine manufacturers' bacon by insisting on this provision. This simple requirement is the reason that there are so many 25-year-old Danish wind turbines still operating in California, while only a few American wind turbines from the period remain in use. Danish wind turbines, while far from perfect, were designed to be failsafe and, for the most part, were.

Wind turbine technology has not progressed so far that we can now safely eliminate this requirement—not even for small wind turbines, where the forces are much less than those on large wind turbines. Even with mechanical or electrical braking and aerodynamic controls, there is still no certainty that a wind turbine

turbines, and the full-span pitch control now common on all large wind turbines (see figure 1-7, Pitchable blade tips.

Some micro turbines, such as Southwest Windpower's Air Breeze, may use electrical or "dynamic" braking as the sole means of control. Others, like the Marlec Rutland 910, may not have any means of braking at all. While this may be tolerable in such small wind turbines, it is not acceptable for larger turbines. Marlec, for example, suggests using the 910 only in areas of moderate winds, as on a sailboat. For land-based applications, where it might encounter destructive winds, Marlec recommends the 910F, the *F* indicating the furling version.

Figure 1-7. Pitchable Blade Tips. Danish wind turbines, like the thousands installed in California during the early 1980s, used centrifugally activated tip brakes to protect the rotor in emergencies. Later designs pitched the blade tips to stop the rotor under normal operation. The blades shown here are a sculpture at the Folkecenter for Renewable Energy in Denmark. The blade on the right has the tip pitched into the emergency stop position. Fail-safe, pitchable blade tips may have been the single most important technological factor in the success of Danish wind turbines and the rise of the commercial wind industry.

Several new micro turbines, such as the Superwind 350 and the Ampair 300, use pitch control to protect the rotor. This is a departure for machines of this size, which have in the past followed the adage, *Keep it as simple as possible* (see figure 1-8, Partial pitch control).

Most mini wind turbines and many household-size turbines "furl" or fold about a hinge so that the rotor swings toward the tail vane. In the past some designs have furled the rotor vertically. Nearly all now furl the rotor hori-

Figure 1-8. Partial Pitch Control. Several new micro turbines, such as this Superwind 350, use partial pitch control to protect the rotor from overspeed. (Superwind, www.superwind.com)

Figure 1-9. Furling Wind Turbine. The Bergey 850 furling toward the tail vane in high winds at the Wulf Test Field.

zontally toward the tail (see figure 1-9, Furling wind turbine). One new household-size wind turbine, however—Southwest Windpower's Skystream 3.7—relies entirely on dynamic braking to control the rotor in high winds. This is a novel approach for a machine of this size, 3.7 meters (12 ft) in diameter.

Some small commercial-scale turbines, such as the Entegrity 50, use tip brakes to protect the rotor, should the mechanical brake fail (see figure 1-10, Tip brakes). Rebuilt Danish wind turbines from California's aging wind farms will use pitchable blade tips that, like tip brakes, deploy only in emergencies.

Figure 1-10. Tip Brakes. Several wind turbines of the 1970s and 1980s used tip brakes to protect the rotor should the brake fail.

Figure 1-11. Full-Span Pitch Control. Pitching blades to feather (into the wind) to control the rotor on a 250 kW WEG MS2 in California's Altamont Pass. The long-discontinued WEG MS2 used a rotor 25 meters (82 ft) in diameter.

Large wind turbines all use pitch control to limit power to the rotor. This wasn't always the case, but as turbines have grown larger, pitching the blades into the wind, or feathering the blades, has become standard. Some products on the market may pitch the blades toward stall, but this has increasingly fallen out of favor (see figure 1-11, Full-span pitch control).

Generators

While wind turbines can be used to mechanically grind grain and pump water—and historically, that's what they were known for—our interest today is in generating electricity. Wind turbines have been used to generate electricity since American inventor Charles Brush's wooden wind dynamo and Poul la Cour's metal windmill at the Askov folk high school in Denmark at the turn of the 19th century. Interestingly, la Cour, the Danish Edison, used electricity to produce hydrogen.[2]

Today most small wind turbines use permanent-magnet alternators. This is the simplest and most robust generator configuration and is nearly ideal for micro and mini wind turbines. Diversity increases with size. For example, Bergey Windpower uses a permanent-magnet alternator in its Excel, a household-size wind turbine, whereas Entegrity uses a conventional induction (asynchronous) generator in its small-commercial scale turbine.

One noticeable characteristic of some permanent-magnet alternators used by small wind turbine manufacturers, such as Bergey Windpower, is their inside-out design. The case to which the magnets are attached, sometimes

Figure 1-12. Permanent-Magnet, Direct-Drive Generator. The Bergey 1.5 kW direct-drive generator. Note that the blades are attached to the "magnet can" (yellow), which rotates outside the stator (red)—a common configuration on many small direct-drive wind turbines.

called the magnet can, rotates outside the stator. In this configuration, the blades can be bolted directly to the case, and they often are (see figure 1-12, Permanent-magnet, direct-drive generator).

Most small wind turbine alternators produce three-phase AC to make best use of the space inside the generator case. Some battery-charging models rectify the AC to DC at the generator; others rectify it at a controller, which can be some distance from the generator.

Though some small wind turbines continue to use inexpensive iron-oxide magnets, many newer small turbines use high performance neodymium-iron-boron or other rare-earth magnets. For example, Southwest Windpower's

---

2. For more on Charles Brush, see Robert Righter's *Wind Energy in America: A History,* University of Oklahoma Press. For more on Poul la Cour, see Paul Gipe's *Wind Energy Comes of Age,* John Wiley & Sons, 1995.

new Skystream uses a purpose-built, directly driven, permanent-magnet alternator that uses neodymium-iron-boron (Ne-Fe-B) magnets.

Small commercial-scale turbines, notably the rebuilt Danish wind turbines from California, typically use off-the-shelf induction generators.

Though some large turbines still use classic induction generators, most have moved to wound-rotor, or what the Europeans call doubly fed, induction generators. These generators allow the wind turbine's rotor to operate at partial variable speed. Variable-speed operation may be crucial in reducing severe loads on the turbine's drive train, thus extending the turbine's life. This is especially true of the very large turbines manufactured today.

Some large wind turbine manufacturers, notably Enercon, build slow-speed generators specifically designed for direct drive. Enercon's large ring generators use electromagnets. Other companies are attempting to use rare-earth permanent magnets.

Wind turbines driving induction generators can be directly connected to the grid without an inverter that converts variable frequency and variable voltage to the constant frequency (60 hertz in the Americas) and constant voltage required by the grid. Alternators and doubly fed induction generators require inverters.

Drive Trains

In nearly all small wind turbines on the market today, the rotor drives the generator directly, without a transmission. As size increases, wind turbines typically require a gearbox to step up

Figure 1-13. Drive Train. Nordex 1 MW nacelle on display at the Husum Messe in Germany. The rotor hub is on the right, the air-cooled generator to the far left, the gearbox in the middle.

the low rpms of the rotor to the higher rpms needed to drive the alternator (see figure 1-13, Drive train). Small commercial-scale turbines and large turbines typically use transmissions of one kind or another. There are important exceptions. Enercon uses a large ring generator that allows the rotor to power the alternator directly. Clipper uses a single-stage transmission to drive multiple, purpose-built permanent-magnet alternators.

Micro Wind Turbines

Micro turbines are the smallest of small turbines and are well suited for very low-power battery-charging applications, such as on sailboats moored in a protected harbor (see table 1-3, Representative Micro Wind Turbines). Southwest Windpower's Air Breeze, which typifies this class of turbine, could be expected to

produce somewhat more than 400 kilowatt-hours per year at a windy site with an average wind speed of 5.5 m/s (12.3 mph).

"Will micro wind turbines generate electricity?" asks Mick Sagrillo rhetorically. Yes, but that doesn't qualify them as "serious" wind turbines, in his estimation. There's a niche market for "recreational wind" turbines, says Sagrillo. These are wind turbines for sailboats, recreational vehicles (RVs), and hobbyists.

There's a difference between using a single solar panel for an RV and using an array of panels to power a home, and the same is true for wind turbines. Micro turbines simply can't produce enough electricity for more power-demanding applications. It doesn't make sense to couple a micro turbine with an inverter in an interconnected application feeding electricity into the grid. The inverter's standby losses alone will eat up the turbine's meager generation, and could lead to a net consump-

Table 1-3: Representative Micro Wind Turbines

Manufacturer	Model		Rotor Diameter		Area	Std. Power Rating*	Mfg. Rated Power	Rated Wind Speed		Perf. at Rated Power	Loading at Rated Power
			m	ft	m²	kW	kW	m/s	mph	%	W/m²
Marlec	Rutland 503	www.marlec.co.uk	0.51	1.7	0.2	0.04	0.03	10.0	22	0.20	122
LVM	Aero4gen F	www.lvm-ltd.com	0.86	2.8	0.6	0.12	0.07	10.3	23	0.18	121
Marlec	Rutland 910-3F	www.marlec.co.uk	0.91	3.0	0.7	0.13	0.09	10.0	22	0.23	138
Ampair	100	www.ampair.com	0.93	3.0	0.7	0.14	0.10	20.0	45	0.03	148
Southwest Windpower	Air Breeze	www.windenergy.com	1.17	3.8	1.1	0.21	0.20	12.5	28	0.16	186
Ampair	300	www.ampair.com	1.20	3.9	1.1	0.23	0.30	12.6	28	0.22	265
Superwind	350	www.superwind.com	1.20	3.9	1.1	0.23	0.35	12.5	28	0.26	309
LVM	Aero6gen F	www.lvm-ltd.com	1.22	4.0	1.2	0.23	0.12	10.3	23	0.15	103
Kestrel	e150	www.kestrelwind.co.za	1.50	4.9	1.8	0.35	0.6	13.5	30	0.23	340

*Standard Power Rating: 200 W/m².

tion of electricity. Still, these can be great little machines at remote sites.

One rugged micro turbine is Marlec Engineering's Rutland 910. The multiblade turbine is ubiquitous in low-power applications in Great Britain, where frequent cloudy skies give the feisty little machine an edge over solar panels (see figure 1-14, Micro wind turbine).

Marlec's Rutland uses a novel pancake generator design that completely seals the generator in plastic. Hugh Piggott calls this an axial flux generator, in contrast with the more common radial field designs used in automotive alternators or magnet-can alternators. Piggott offers a peek inside the Marlec in his book *Windpower Workshop*, and describes how the unusual generator works.

If you see a small multiblade turbine in a yacht basin and it's not a Marlec, it's most likely an Ampair. This hardy little machine uses a more conventional configuration than the Marlec. The venerable Ampair 100 was the only wind turbine at the Wulf Test Field that met its manufacturer's advertised power curve.

Southwest Windpower put sex appeal into wind energy with the introduction of its sleek Air series of micro turbines. Priced low to stimulate volume sales, the Air series was introduced at about half the price per rotor area of competitive products. The combination of price and visual appeal worked, and Air turbines have found wide application.

Early versions of the Air, models 303 and 403, were intended as an off-the-shelf consumer commodity to complement photovoltaic panels in hybrid systems. Unfortunately, the lightweight, high-tech design wasn't ready for market when it was released; the unit developed

Figure 1-14. Micro Wind Turbine. David Nixon displays the "sun flower," which sports solar panels and a Rutland 910 micro wind turbine, at the Kortright Centre for Conservation, Ontario, Canada.

a reputation for having poor reliability and being particularly noisy. Later versions, such as the Air Breeze, are far more reliable and significantly quieter.

Mini Wind Turbines

Mini wind turbines capture from 2 to 10 times more energy than micro turbines, making them suitable for small cabins at remote sites. If the Air Breeze in our example produces about 400 kWh per year at a 5.5 m/s (12.3 mph) site, then mini turbines can produce from 800 kWh to 4,000 kWh per year (see table 1-4, Representative Mini Wind Turbines).

Table 1-4: Representative Mini Wind Turbines			Rotor Diameter		Area	Std. Power Rating*	Mfg. Rated Power	Rated Wind Speed		Perf. at Rated Power	Loading at Rated Power
Manufacturer	Model		m	ft	m²	kW	kW	m/s	mph	%	W/m²
Kestrel	e150	www.kestrelwind.co.za	1.5	4.9	1.8	0.35	0.6	13.5	30	0.23	340
Ampair	600	www.ampair.com	1.7	5.6	2.3	0.45	0.90	10.0	22	0.65	397
Windsave	WS 1000	www.windsave.com	1.8	5.7	2.4	0.48	1.00	12.0	27	0.39	416
Zephyr Corporation	Airdolphin	www.zephyreco.co.jp	1.8	5.9	2.5	0.51	1.00	12.5	28	0.33	393
Marlec	Rutland 1803-2	www.marlec.co.uk	1.8	6.0	2.6	0.52	0.34	10.0	22	0.21	130
J. Bornay	Inclin 600	www.bornay.com	2.0	6.6	3.1	0.63	0.60	11.0	25	0.23	191
Renewable Devices	Swift	www.renewabledevices.com	2.1	6.9	3.5	0.69	1.50	12.5	28	0.36	433
Southwest Windpower	Whisper 100	www.windenergy.com	2.1	7.0	3.6	0.72	0.90	10.0	22	0.41	252
Fortis	Espada	www.fortiswindenergy.com	2.2	7.2	3.8	0.76	0.60	12.0	27	0.15	158
Bergey Windpower	XL.1	www.bergey.com	2.5	8.2	4.9	0.98	1.00	11.0	25	0.25	204
Southwest Windpower	Whisper 200	www.windenergy.com	2.7	8.9	5.7	1.15	1.00	10.5	24	0.25	175
Proven Wind Turbines	2.5	www.provenenergy.com	3.5	11.5	9.6	1.92	2.50	10.0	22	0.42	260

*Standard Power Rating: 200 W/m².

Figure 1-15. Mini Wind Turbine. The Bergey XL.1 uses a rotor 2.5 meters (8.2 ft) in diameter and sweeps nearly 5 m² of the wind stream. (Detronics, www.detronics.net)

Most mini wind turbines are best suited for battery applications. Some of the larger turbines in this class, such as Bergey Windpower's XL.1, can be found in hybrid systems powering off-the-grid households. Several Bergey XL.1s and Southwest Windpower Whisper 100s, for example, can be seen powering off-the-grid homes in Tehachapi's Mountain Meadows subdivision (see figure 1-15, Mini wind turbine).

Some mini wind turbines, such as Renewable Devices' Swift and Windsave's WS1000, have been mounted on rooftops in urban areas in Great Britain. The turbines were then connected to the grid with inverters. This is a misapplication of wind technology. The turbines would be better suited elsewhere. They would work better and produce more electricity if mounted on a tall tower so that they don't have to struggle through turbulent rooftop winds.

Household-Size Wind Turbines

Household-size wind turbines are physically big machines (see table 1-5, Representative Household-Size Wind Turbines). For example, at one end of the scale is Abundant Renewable Energy's 110. It sweeps 10 m²—10 times the

Table 1-5: Representative Household-Size Wind Turbines			Rotor Diameter		Area	Std. Power Rating*	Mfg. Rated Power	Rated Wind Speed		Perf. at Rated Power	Loading at Rated Power
Manufacturer	Model		m	ft	m²	kW	kW	m/s	mph	%	W/m²
Abundant Renewable Energy	110	www.abundantre.com	3.6	12	10.2	2.0	2.5	11.0	25	0.30	246
Westwind	3	www.westwindturbines.co.uk	3.7	12	10.8	2.2	3.0	14.0	31	0.17	279
Southwest Windpower	Skystream	www.windenergy.com	3.7	12	10.9	2.2	1.9	8.9	20	0.40	175
Kestrel	e400	www.kestrelwind.co.za	4.0	13	12.6	2.5	3.0	12.5	28	0.20	239
Southwest Windpower	Whisper 500	www.windenergy.com	4.6	15	16.4	3.3	2.4	10.0	22	0.24	146
Fortis	Montana	www.fortiswindenergy.com	5.0	16	19.6	3.9	4.2	16.0	36	0.09	214
SMA	Aerosmart	www.aerosmart.de	5.1	17	20.4	4.1	5.0	13.5	30	0.16	245
Westwind	5	www.westwindturbines.co.uk	5.1	17	20.4	4.1	5.0	14.0	31	0.15	245
Proven Wind Turbines	6	www.provenenergy.com	5.5	18	23.8	4.8	6.0	11.5	26	0.27	253
Endurance Wind Power	S250	www.endurancewindpower.com	5.5	18	23.8	4.8	5.0	—	—	—	210
Eoltec	Scirocco	www.eoltec.com	5.6	18	24.6	4.9	6.0	11.5	26	0.26	244
Westwind	10	www.westwindturbines.co.uk	6.2	20	30.2	6.0	10.0	14.0	31	0.20	331
Bergey Windpower	Excel-S	www.bergey.com	6.7	22	35.3	7.1	10.0	13.8	31	0.17	284
Aircon	10	www.aircon-international.com	7.1	23	39.6	7.9	9.8	11.0	25	0.30	248
Abundant Renewable Energy	442	www.abundantre.com	7.2	24	40.7	8.1	10.0	11.0	25	0.30	246

*Standard Power Rating: 200 W/m².

sweep of the Air Breeze—and weighs more than 140 kilograms (300 lbs). At the other end of the scale is ARE's 442. It intercepts more than 40 m² and weighs more than 600 kilograms (~1,400 lbs). All else being equal, the ARE442 should be able to capture 40 times more wind energy than the Air Breeze.

For comparison, then, at a 5.5 m/s (12.3 mph) site, ARE's 110 could produce 10 times as much electricity as the Air Breeze, or somewhat more than 4,000 kWh per year. Abundant Renewable Energy estimates that the 110 will generate 4,800 kWh per year, about 20 percent more than the back-of-the-envelope method suggests. There could be good reasons for better performance than calculations based on the average performance of machines in that class. The ARE110 uses high-flux density, rare-earth magnets (Ne-Fe-B) that are superior to less expensive but more common ferrite magnets. As the mileage placards on new cars warn, "Your performance may vary." (See figure 1-16, Household-size wind turbine.)

Small Commercial-Scale Wind Turbines

There are few products in the small commercial-scale class of wind turbines (see table 1-6, Representative Small Commercial Wind Turbines). Enercon currently does not sell turbines in the United States, though it does install turbines in Canada. Fuhrländer has no more than a handful of turbines in North America, and Gaia has none. The Prince Edward Island manufacturer Entegrity is the only company actively marketing turbines in this size class in North America. Its EW50 is a modern version of a turbine developed in the 1980s. However, they have incorporated the lessons learned from those early machines, notably correcting problems with the Enertech's troublesome tip brakes.

Swept Area Rules of Thumb

There are some simple rules of thumb to remember when comparing the size of different wind turbines in terms of their potential power output. A micro turbine that sweeps 1 m² is roughly equivalent to a power rating of 200 watts. A mini wind turbine that sweeps 10 m² would be rated at about 2 kW. A household-size wind turbine that sweeps 100 m² would be rated from 25 to 40 kW. And a large wind turbine that sweeps 5,000 m² could be rated from 1.5 to 2.5 MW.

Table 1-sb1: Swept Area, Rotor Diameter, and Nominal Power Rules of Thumb

Swept Area m²	Nominal ft²	Nominal Rotor Diameter m	Nominal ft	Nominal Power Rating kW
1	10	1.1	4	0.2
5	50	2.5	8	1
10	110	3.6	12	2
50	540	8	26	10–20
100	1,080	11	37	25–40
1,000	10,800	36	118	300–400
5,000	53,800	80	262	1500–2500

Figure 1-16. Household-Size Wind Turbine. The Bergey Excel at a home in New England. The Excel, a 7 m (22 ft) diameter wind turbine, uses a direct-drive, permanent-magnet generator. The rugged wind turbine has been the workhorse of the household-size wind turbine market for three decades.

Table 1-6: Representative Small Commercial Wind Turbines			Rotor Diameter		Area	Mfg. Rated Power	Rated Wind Speed		Perf. at Rated Power	Loading at Rated Power
Manufacturer	Model		m	ft	m2	kW	m/s	mph	%	W/m²
Enercon	E12	www.enercon.de	12	39	113	30	11.0	25	0.33	265
Fuhrländer	F30	www.fuhrlaender.de	13	43	133	30	12.0	27	0.21	226
Gaia Wind	11 kW	www.gaia-wind.com	13	43	133	11	9.5	21	0.16	83
Entegrity	EW 15	www.entegritywind.com	15	49	177	50	13.0	29	0.21	283
PGE	20/35	www.energiepge.com	19	63	290	35	11.0	25	0.15	121

There are several firms buying, reconditioning, and marketing used turbines from California wind farms. Most of these are rebuilt Danish wind turbines. Names to look for are Vestas, Windmatic, Bonus, and Nordtank—in that order. However, some used California turbines are "Rust-Oleum rebuilts," as Mick Sagrillo disparagingly calls them. This refers to the slapdash repainting of early American-designed machines, including the Enertech E44. The E44 is similar to Entegrity's EW50. However, they are not the same machine. Avoid the E44, and avoid the Carter, StormMaster, and ESI brands (see figure 1-17, Small commercial-scale wind turbine).

Some companies are also importing used Danish wind turbines to the North American market. One example is the Vestas V27 installed in Cleveland at the Great Lakes Science Center. Bear in mind that used turbines imported from Europe must be converted from 50-hertz AC to the 60-hertz AC used in the Americas. This is not as simple as it sounds, and some buyers in Canada have found out the hard way that buying a used turbine from Europe can become a nightmare.

Figure 1-17. Small Commercial-Scale Wind Turbine. A used Bonus wind turbine from a California wind farm is operated by a rancher near Pincher Creek, Alberta. In the early 1990s several used Danish wind turbines were installed in southwestern Alberta.

Large Wind Turbines

Large or commercial-scale wind turbines are the class of machines being used in the giant wind power plants springing to life across the continent. Demand is so great that few turbines turbines, they convert far more of the energy in the wind to electricity than, say, the Air Breeze. As Scottish small wind turbine designer Hugh Piggott notes, in wind energy "size matters."

Most manufacturers of large wind turbines are European and Asian, though GE Wind also

As Scottish small wind turbine designer Hugh Piggott notes, in wind energy "size matters."

are actually available as this book is going to press. Many of the big wind farm developers have orders for hundreds of these giant machines many years into the future (see table 1-7, Representative Large Wind Turbines).

These turbines are so large that they can effectively intercept 5,000 times more area of the wind stream than an Air Breeze micro turbine. And because these large wind turbines are so much more efficient than small wind

has plants in North America, and Clipper has started ramping up assembly in Iowa.

Not all large wind turbines on the international market are available in the United States because of restrictive trade practices. Previously, the models available in North America have not been the state-of-the-art ones found in Europe. This has begun to change, notably in Canada. Enercon, one of the world's leading manufacturers, does not sell to the American market,

Table 1-7: Representative Large Wind Turbines

Manufacturer	Model		Rotor Diameter		Area	Mfg. Rated Power	Rated Wind Speed		Perf. at Rated Power	Loading at Rated Power
			m	ft	m²	kW	m/s	mph	%	W/m²
GE	1.5sl	www.gepower.com	77	253	4,657	1,500	12	27	0.30	322
Fuhrländer	MD 77	www.fuhrlaender.de	77	253	4,657	1,500	12	26	0.34	322
Ecotecnia	80	www.ecotecnia.com	81	264	5,090	2,000	13	29	0.29	393
Vestas	V82	www.vestas.com	82	269	5,281	1,650	13	29	0.23	312
Siemens	2.3-82	www.powergeneration.siemens.com	82	269	5,281	2,300	13	29	0.32	436
Enercon	E82	www.enercon.de	82	269	5,281	2,000	12	27	0.36	379
Suzlon	S88	www.suzlon.com	88	289	6,082	2,100	14	31	0.21	345
Nordex	N90	www.nordex-online.com	90	295	6,362	2,500	14	31	0.23	393
Gamesa	G90	www.gamesa.es	90	295	6,362	2,000	17	38	0.10	314
Clipper	2.5	www.clipperwind.com	93	305	6,793	2,500	13	29	0.27	368

Figure 1-18. Large Wind Turbine. Several GE 1.5 MW wind turbines near Montfort, Wisconsin. These turbines use a rotor 77 meters (250 ft) in diameter.

but its turbines are available elsewhere in the Americas (see figure 1-18, Large wind turbine).

Danish manufacturer Vestas, German producer Siemens (Bonus), Spanish company Gamesa, and of course GE are the main wind turbine suppliers for wind power plants in the United States. Indian manufacturer Suzlon has installed a significant number of machines in the upper Midwest.

Vertical-Axis Revival

Since *Wind Energy Basics* was first published in 1999, there has been an explosion of interest in new vertical-axis wind turbines. In that edition, the conclusion illustrating one photograph was stark—"Practically no one is working with vertical axis wind turbines today." The text went even farther when describing the difficulties that "new" or "novel" designs face:

Even promising designs from legitimate manufacturers encounter enough technical and financial problems that only a few survive. Darrieus or "eggbeater" turbines are one example. Sleek and relatively simple, they promised lower costs and greater reliability than conventional wind turbines. But they didn't deliver.

Darrieus designs failed first in the demanding small turbine market. They couldn't compete. Then they failed in the commercial or wind farm market. Today only a few Darrieus turbines remain standing, and soon these will all be removed for scrap.

Harsh? Yes. It was a conclusion based on bitter disappointment in the Darrieus wind turbines of the day, especially the derelict Darrieus turbines once littering California's windy passes.

Have times—and technology—changed? Possibly. That interest and experimentation in

if not generations. The exceptions are the few Darrieus turbines, such as the Turby, the Quiet Revolution, and the MARC-Twister, that use curvilinear blades. These designs have not been seen before. Whether they are revolutionary or not, or whether they are better than preceding Darrieus designs, only time and field experience will tell.

Another of the "show me it works" skeptics is Mick Sagrillo. And when he says "show me it works," he means show him that it produces electricity in kilowatt-hours—not just instantaneous power. After all, it is wind energy in kilowatt-hours that we're after, not power in kilowatts.

For whatever reason, many VAWT designers are more prone to hyperbole than most other wind turbine designers.

---

new Darrieus technology is high can be clearly seen in the dozens of Web sites devoted to various vertical-axis designs. Why this is so is much less clear.

Not everyone is thrilled at the VAWT renaissance. "The driving force [behind the revival of VAWTs]," says Ian Woofenden, "is ignorance of past failures, and arrogance about overcoming the problems" inherent in the designs. Woofenden, an editor who lives with conventional small wind turbines off the grid, points out that VAWTs are not new. He admonishes that "anyone designing a new one [VAWT] should do their homework first to find out what past designers learned."

As Woofenden notes, few of the "new" designs are in fact new at all. Many of the designs touted as new have been around for decades,

Sagrillo grumbles that videos of VAWTs "spinning" can be found on YouTube.com. One famously showed men installing a purported wind turbine on the roof of a building to dramatic music; immediately at the climax the "turbine" started spinning. Indeed, *spinning* is the operative word, because the rotor of this "turbine" wasn't connected to anything: no generator, no gearbox, no drive train at all. People are "dazzled by the motion," says Sagrillo in despair. Most viewers never notice that the "turbines" are not doing anything productive, only spinning. As Sagrillo pithily notes, "If that's all you want, buy a whirligig for $12 and be done with it."

VAWT skeptics, such as Woofenden and Sagrillo, hold inventors of conventional wind turbines to the same standards as they do vertical-axis designers. They don't single out vertical-axis designs solely because they use a vertical axis, as some VAWT proponents charge. As Jon Powers, a Tehachapi windsmith with decades of operational experience on conventional wind turbines, notes, if inventors develop a nontraditional wind turbine, they should prove it, commercialize it, and deploy it by the thousands. Then it, too, will become the new conventional thinking. And, he could add, everyone, skeptics included, would applaud their success.

For whatever reason, many VAWT designers are more prone to hyperbole than most other wind turbine designers. Some claim their turbines will produce at less cost and with less impact on the environment than conventional wind turbines. Maybe such claims are due to widespread ignorance of VAWT technology or its long history. Maybe vertical-axis designers imagine that their turbines don't suffer the limitations of "all those other ordinary wind turbines." Certainly it's easier to visualize how VAWTs, like cup anemometers, work than how those thin spindly blades on conventional wind turbines can be so efficient at extracting the energy in the wind.

Do VAWTs work? Of course they do. This isn't in question. Can they produce usable quantities of electricity? Yes, the record in California is clear: Darrieus wind turbines generated millions of kilowatt-hours for nearly a decade. Can they compete with conventional wind turbines? Perhaps. Are there specialized markets for which VAWTs may be ideally suited? Possibly. But by definition, a specialized market is a limited market. Do we need wind turbines for specialized markets? Yes, we'll need all the renewable energy we can find in the coming decades. Do VAWTs offer a panacea of limitless renewable energy without the drawbacks of conventional wind turbines? Unlikely.

VAWT proponents sometimes claim there's little or no evidence that their designs won't do what they promise. Maybe so, but that begs the question. The burden of proof lies with the inventors to prove that their claims are true. As critics of newfangled inventions point out, it's impossible to prove that they won't work. Students in Logic 101 learn that you can't prove a negative. There's always a reason, an excuse really, why many new, earth-saving inventions don't work quite as well as advertised and are then quickly forgotten.

Here, then, are some of the arguments VAWT proponents use to contrast their technology with that of conventional wind turbines. Note that the arguments and the following responses can be applied equally well to any proposed wind turbine that is advertised as a technological breakthrough.

- ***They are simpler.*** Yes, some VAWTs are simpler than conventional wind turbines. Others, however, are more complex. In the end, VAWT designs often trade one form of complexity for another.
- ***They are more reliable.*** Possibly, but unlikely. In nearly all cases, proponents have no field experience to support such a claim. Often the claim is based only on a wish, not on real performance. With the exception of Windside, the Finnish manufacturer of S-rotors, few

manufacturers of VAWTs today have any operating experience. None has performance data in the public domain where independent analysts can gauge the reliability of their design.

- **They are less costly.** They may indeed be cheaper than conventional wind turbines. But if the turbine doesn't work at all, doesn't work well, or doesn't work for long, it's no bargain. In one North American case, the turbine was far more costly than the conventional wind turbines it was said to replace.

- **They are more efficient.** Calculation of efficiency from wind tunnel or truck tests says very little about how a wind turbine will operate in real winds. Even where a wind turbine is markedly more efficient than another, reliability is a far more critical parameter. A wind turbine may be efficient, but if it is not working reliably, it will produce little or no electricity.

- **They are more cost-effective.** Whether a wind turbine is a good buy or not is a function of its installed cost, the amount of energy it generates, and the cost of operating and maintaining it. Modern VAWTs may be a good buy relative to other wind turbines, but there is so little data on their performance that no one can say for sure.

- **They don't kill as many birds.** This is simply an unfounded claim. Very few studies have been done on birds and modern VAWTs because there are so few VAWTs in operation. Nearly all studies to date have been conducted on large commercial wind turbines. These studies find that the number of birds killed by wind turbines is primarily a function of turbine size. A big wind turbine will kill proportionally more birds than a small wind turbine— of any configuration. Proponents argue that there's no evidence that modern VAWTs kill birds at all. While technically true, the reason for this paradox is that there are no studies on modern VAWTs and birds. Moreover, nearly all modern VAWTs are extremely small, and the likelihood of a small wind turbine— of any configuration—killing any bird is, therefore, very small.

- **They are less noisy.** This is one claim that may have real merit. The blades on modern VAWTs may move through the air at much lower speeds than blades on conventional wind turbines. The lower blade speeds often translates into lower noise emissions than those from conventional wind turbines. Unfortunately, there is very little field experience, and even less publicly available data, to verify this assertion.

VAWT design, like the design of any other wind turbine, is a series of trade-offs. For each

Fantasy Wind Turbines; or, If It's Too Good to Be True . . . : How to Spot Scams, Frauds, and Flakes

German engineering professor Robert Gasch calls them fantasy wind turbines. These are the inventions—or contraptions—that bedevil serious wind turbine advocates. They are the "revolutionary" inventions that periodically rise up from the dead whenever the price of oil goes up or there's a "power crisis" somewhere in the world.

Very few of these ideas are new, and certainly none of them is revolutionary. While some of these devices may spring from well-intentioned inventors, others are the conceptions of fast-buck artists. David Sharman, a serious wind turbine designer, calls the proponents of such machines "bozos and shysters."

It's often difficult for the uninitiated to tell the difference between the real and the imaginary, the fraudulent from the worthwhile—and that's the problem.

Here are some tips for spotting questionable products. The most important tip to keep in mind comes from Professor Gasch: If there is a new wind turbine, no one should pay the slightest attention to it until they "build it, measure it, and publish" the results. Until then, it's just hot air—marketing hype—and nothing more.

How can you identify a questionable wind turbine design? Here are some tips.

- Hype high, experience low (or nonexistent).
- Aggressive marketing. Look for multilevel or pyramid schemes of the *Get in on the ground floor now* variety.
- New design: It's "not like the others."
- "New" patents.
- Drag devices (squirrels in a cage).
- Ducted turbines.
- No hardware, but a fancy Web site. Web sites are always cheaper than hardware.
- "Works in low wind."
- Is silent. (No wind turbine is silent.)
- Does not kill birds. (All wind turbines can kill birds or bats.)

What lessons have we learned from 30 years of modern wind turbine development?

- There are no panaceas.
- There are no cheap solutions.
- There are no breakthroughs—no miracles.
- Numbers matter. (Wind energy is always about numbers.)
- Experience matters. (If they haven't been building these things for years, then how do you know that it works?)
- Size matters. (You can't get a lot of electricity from a small rotor.)

In sum, always be wary of "new" designs. There's rarely anything truly new under the sun—or in the wind—and "if it seems too good to be true, it probably is."

plus there is a minus. We can't evaluate design elements in isolation. We must consider the wind turbine as a complete package.

For example, the lower blade speed of modern VAWTs is a design element, a by-product of which is lower noise emissions. However, Jim Tangler, a retired aeronautical engineer, notes that VAWTs derive more of their power from torque than do conventional wind turbines because they spin at lower speeds. To handle the greater torque, the blades and their supports must be stronger. This results in greater mass, and hence often greater cost than for similar components of a conventional wind turbine. Further, says Tangler, the blades of VAWTs only operate at optimum aerodynamic performance over a small portion of their carousel path, and they typically use less efficient symmetrical airfoils than the cambered (asymmetrical) airfoils found on conventional wind turbines. All in all, says Tangler, VAWTs of the 1980s weighed more and were less efficient than conventional turbines of the period.

That said, VAWTs could be cost-effective—and even more so than conventional wind turbines—if they were cheap and reliable enough. That is ultimately the test, the Holy Grail for modern VAWTs—or wind turbines of any stripe.

The World's Most Successful VAWT

The most successful VAWT in history was that developed by FloWind in the early 1980s. Using what has become a rather conventional two-blade phi configuration or "eggbeater" Darrieus, FloWind installed more than 500 turbines in California's Altamont and Tehachapi passes (see figure 1-19, FloWind Darrieus).

By the end of 1985, FloWind had installed

95 megawatts of its signature product. For comparison, that's equivalent to more than 15,000 of Quiet Revolution's sleek 5-meter (16 ft) diameter turbine.

Figure 1-19. FloWind Darrieus. One of the hundreds of FloWind Darrieus turbines installed atop Cameron Ridge in the Tehachapi Pass during the early 1980s. These turbines were the most successful VAWTs ever built and operated into the mid-1990s. None remain.

At its most productive in 1987, FloWind's fleet generated 100 million kilowatt-hours—enough electricity for nearly 20,000 California homes. No VAWT manufacturer to this day has ever come close to rivaling that accomplishment. On any windy day in Tehachapi during the late 1980s, the sun could be seen glinting off hundreds of FloWind's turbines spinning atop Cameron Ridge.

Generation began collapsing, however, as serial failures in the joints between the sections of extruded aluminum blades overwhelmed the American company. By the late 1990s FloWind was generating one-tenth the electricity it had in 1987. Eventually all of its Tehachapi turbines were removed and sold for scrap. Some relics were still standing in the Altamont Pass in 2003, though they, too, were slated for removal.

The wind industry learned a lot from the experience. FloWind proved without a doubt that VAWTs could reliably generate commercial quantities of electricity—at least for a decade. But the firm's aggressive marketing and high power ratings have tainted vertical-axis technology ever since. Much of today's cynicism about new VAWTs derives from FloWind's hype about its turbines and from its manipulation of power ratings.

FloWind's turbines were large commercial wind turbines of the day. The company built two models: a 17-meter (56 ft) and a 19-meter (62 ft) version. The 17-meter turbine, for example, was about 17 times the size of the architecturally dramatic Quiet Revolution Q5.

Characteristic of FloWind's marketing, and that of other Darrieus turbines of the day, including DAF-Indal and Vawtpower, was the turbines' high power ratings. FloWind's 17-meter model was rated at 142 kW at a wind speed of 38 mph (17 m/s), while its 19-meter model was rated at 250 kW at a wind speed of 44 mph (~20 m/s). For comparison, that's a Force 8 or "fresh gale" on the Beaufort scale. Translation: That's so windy, no one in their right mind wants to be outside.

To understand these power ratings, it's necessary to look at the area swept by the eggbeater-shaped rotor. FloWind's 17-meter model swept 260 m^2, equivalent to a conventional wind turbine 18 meters (60 ft) in diameter. The 19-meter model swept 340 m^2, equivalent to a conventional wind turbine 21 meters (70 ft) in diameter. Today turbines of this size are considered small commercial turbines. For comparison, a conventional wind turbine 18 meters in diameter would typically be rated at 100 kW, and a 21-meter turbine would be rated at 150 kW.

The high ratings of the FloWind machines translate into a specific capacity of 546 W/m^2 for the 17-meter model and an incredible 735 W/m^2 for the 19-meter model. That FloWind was greatly overstating the potential performance of its turbine was reflected in its average capacity factor, a measure of performance relative to the size of the turbine's generator. The capacity factor of FloWind's turbines never exceeded 12 percent on average and was often less than 10 percent at a time when conventional wind turbines were delivering twice that.

Why were FloWind's power ratings so high? Wind turbines of that era were often sold to uninformed investors who compared wind turbine prices based on the cost per kilowatt of installed capacity. FloWind's aggressive power ratings enabled the company to charge far more

Specific Capacity and Specific Yield

Because there is no standard rating system for wind turbines, manufacturers are free to "rate" their products at whatever wind speed they desire. This leads to confusion, even among professionals. There is simply no way to directly compare the "power rating" of one wind turbine with that of another unless the "rated wind speeds" are identical.

As explained elsewhere, the most important descriptor of a wind turbine's size is the area swept by its rotor in square meters (m²). Consequently, the simplest way to compare the "rated power" of one wind turbine to another is by examining its specific capacity or "rotor loading" in watts per square meter (W/m²). By using this measure, you can quickly compare claims between manufacturers. For example, the Standard Power Rating used in *Wind Energy Basics* is based on 200 W/m². Any claim that a small wind turbine can greatly exceed 200 W/m² should be viewed with a good deal of skepticism. However, large wind turbines can deliver specific capacities of as much as 400 W/m².

The measure of annual performance used worldwide by the wind industry is specific yield in kilowatt-hours per square meter per year (kWh/m²/yr). While some in North America continue to use "capacity factor," this is an archaic and misleading indicator of performance. Modern large wind turbines at windy sites will produce about 1,000 kWh/m²/yr. Small wind turbines will—at best—produce only half that under the same conditions.

for its turbines than they were worth. FloWind's turbines were never truly in the 150 kW or 250 kW size class, but that's what investors were told.

Despite these outlandish power ratings, FloWind's Darrieus designs turned in a respectable performance relative to the area of the wind stream swept by the two-blade rotor. During good years, the machines would generate 500 to 600 kWh/m²/year, competitive with conventional wind turbines of the day.

The bottom line: FloWind's Darrieus turbines operated for about a decade generating millions of kilowatt-hours, and in doing so delivered respectable performance until fatigue and design weaknesses led to increasing unreliability and they were removed. FloWind's turbines, when in regular service, delivered about the same performance as nearby conventional wind turbines relative to their swept area, but performed poorly in comparison with their inflated power ratings.

Because of scale effects, it is unlikely that the small wind turbines of the modern VAWT revival will approach the historical performance of FloWind's large wind turbines.

Modern Small VAWTs

Many manufacturers of the small VAWTs of the vertical-axis revival, like FloWind before them, rate the power output of their turbines at a much higher level than conventional wind turbines. This can be seen in the rotor loading of modern VAWTs that are about double that of conventional turbines (see table 1-8, Representative Vertical-Axis Wind Turbines).

The characteristic loading of conventional small wind turbines is about 250 W/m². For

Table 1-8: Representative Vertical-Axis Wind Turbines

Manufacturer	Model		Rotor Height m	Rotor Height ft	Rotor Diameter m	Rotor Diameter ft	Swept Area m²	Std. Power Rating* kW	Mfg. Rated Power kW	Rated Speed m/s	Rated Speed mph	Perf. At Rated Power %	Loading at Rated Power W/m²
Vertica	3kW	www.vertica-inc.com	1.2	3.9	3.0	4.7	3.6	0.72	3.00	—	—	—	833
Marc Power	Twister 1000	www.marcpower.com	1.9	6.2	1.9	6.2	3.6	0.72	1.00	12.0	27	0.26	277
Mariah Power	Windspire	www.mariahpower.com	6.0	19.7	0.6	2.0	3.7	0.73	1.00	11.0	25	0.34	273
PacWind	Delta I	www.pacwind.net	2.0	6.5	2.0	6.6	4.0	0.80	2.00	12.5	28	0.42	503
Windside Oy	4	www.windside.com	4.0	13.1	1.0	3.3	4.0	0.80	0.24	15.0	34	0.03	60
Turby		www.turby.nl	2.0	6.6	2.7	8.7	5.3	1.06	2.50	14.0	31	0.28	472
Wind Terra	1	www.windterra.com	2.3	7.4	2.7	8.7	6.0	1.20	1.00	11.0	25	0.20	167
Ropatec	Simply Vertical	www.ropatec.com	3.3	10.8	2.0	6.6	6.6	1.32	3.00	14.0	31	0.27	455
Cleanfield	3.5	www.cleanfield.energy.com	2.8	9.0	3.0	9.8	8.3	1.65	3.50	12.5	28	0.35	424
Quiet Revolution	Q5	quietrevolution.co.uk	5.0	16.4	3.1	10.2	15.5	3.10	6.00	12.0	27	0.37	387

*Std. Power Rating for micro, mini, and household-sized wind turbines = swept area x 200 W/m²

example, the Bergey Excel has a rotor loading of 260 W/m² while that of Southwest Windpower's Skystream is 166 W/m². Southwest Windpower's Air 403 was notorious for its high rotor loading of 373 W/m². Fortunately, the firm cut the rating of the Air Breeze, the newest model in the Air series, to only 186 W/m², a much more realistic value.

The high rotor loading of many modern VAWTs doesn't mean that the wind turbine is capable of generating more electricity than a conventional wind turbine of the same swept area. It simply means that the wind turbine uses a much larger generator relative to the area swept by the wind turbine rotor. Remember

that it's the swept area that is the prime determinant of how much energy a wind turbine will capture—not the size of the generator or its "power" rating.

High rotor loading does, however, suggest that a VAWT manufacturer may be overstating expected performance. Rotor loading greater than 300 W/m² for small wind turbines should be viewed with skepticism, whether on the overhyped Air 403 or on a small VAWT.

Another way to look at rotor loading is to consider a Standard Rated Power of 200 W/m². Many—though certainly not all—VAWTs in table 1-8 have rated their turbines at twice the Standard Power Rating. PacWind, a new straight-blade VAWT touted by Hollywood celebrities, has a power rating of 500 W/m². PacWind rates the Delta I at 2 kilowatts, when in all likelihood it is less than a 1 kW turbine. Worse yet is Vertica, which rates its VAWT at 3 kilowatts, or more than 800 W/m² for a turbine that, like PacWind's Delta I, is more realistically labeled something less than 1 kilowatt.

Note that the performance at rated power and the projected loading at rated power in the table are based on manufacturers' advertised characteristics. That a manufacturer claims a conversion efficiency of 30 percent and rotor loading of anything greater than 200 W/m² does not guarantee that the wind turbine will actually deliver such performance. In fact, very few small wind turbines—of any configuration—have reliably delivered such performance.

Small VAWT Designs

During the 1970s and early 1980s, several designs for vertical-axis wind turbines with straight blades emerged. These H-rotors used articulating blades that changed pitch as they moved around the carousel path (see figure 1-20, Articulating, straight-blade VAWT). Articulating H-rotors never performed as advertised and were mechanically unreliable.

Figure 1-20. Articulating, Straight-Blade VAWT. The Pinson Cycloturbine used a wind vane on top of the turbine to control the pitch of the articulating blades as the rotor turned in the wind. This unit was installed at a technical college near Las Cruces, New Mexico, in 1979.

Figure 1-21. Nonarticulating, Straight-Blade VAWT. Cleanfield's experimental VAWT on the rooftop of a technical college near Hamilton, Ontario. (Martin Ince, www. mkince.ca)

The recent VAWT revival has seen the introduction of several new small H-rotors with blades that are fixed in pitch like those of Darrieus turbines. Two examples are turbines made by the Canadian company Cleanfield, and California's PacWind (see figure 1-21, Nonarticulating, straight-blade VAWT). In contrast with earlier articulating VAWTs, these turbines are simpler. Unfortunately, there is little to no information on how these turbines perform in the field.

As with conventional wind turbines, new vertical-axis designs should have some aerodynamic means of protecting the rotor should the normal braking system fail. Some Darrieus turbines of the 1970s and 1980s used air brakes; others (FloWind) did not, relying solely on their mechanical brakes (see figure 1-22, Old Darrieus with air brakes).

Novel Turbines

There is a never-ending stream of novel wind turbine designs. Many are reinventions of past designs that have been forgotten, or variations on a common theme, of which folding buckets and articulating paddles are typical examples. Yet some, such as Doug Selsam's SuperTwin, are truly novel.

Figure 1-22. Old Darrieus with Air Brakes. A DAF-Indal Darrieus wind turbine at the Kortright Centre for Conservation. DAF-Indal designs characteristically used twin centrifugally activated air brakes (flaps) at the midpoint of each blade. The air brakes were intended to protect the rotor should the brake fail. This derelict turbine has stood idle for nearly two decades and should be in a museum somewhere. Note that there is no drive train beneath the rotor.

Figure 1-23. Doug Selsam's Experimental SuperTwin. If one rotor is good, two are better, believes the inventor. Here the performance of the SuperTwin is being measured at the Wulf Test Field in windy Tehachapi Pass.

How to Evaluate New Small Wind Turbine Technology

Regardless of the official status of standards and certification, there is an international consensus that the only way to properly test small wind turbines is in the field. Both the American Wind Energy Association's performance testing standard and the proposed European standard for small wind turbines exclude performance measurements of wind turbines in wind tunnels or through truck tests, because neither "truly replicates free-stream conditions." Worse, wind tunnel tests typically overstate real-world performance. This phenomenon was seen in the 1970s during development of government-sponsored wind turbines, notably the US development of McDonnell Aircraft's giromill. Typically, consumers will never see the performance measured in either wind tunnel or truck tests.

The absolute minimum a consumer should demand of any manufacturer are the results of performance tests conducted in accord with standard international practice. This information should be available either in product literature or on the manufacturer's Web site. The manufacturer should clearly state whether the tests were done to the international standard. The resulting data can be presented as a power curve or as a curve of estimated annual energy production (AEP) or sometimes the annual energy output (AEO). Most importantly, the data must be averaged over a period of time—10 minutes for large wind turbines, and 1 minute for small wind turbines.

Ultimately, the small turbine industry will move toward standardized performance labeling. While the power curve of average data is useful, the most important function of labeling is to report the amount of energy in kWh that the wind turbine will produce at, say, an average annual wind speed of 5 m/s (~11 mph). The labeling of standardized noise measurements will eventually appear as well—when consumers demand it.

Selsam's SuperTwin

"If one rotor is good, two are better," says the inventor of the SuperTwin (see figure 1-23, Selsam's SuperTwin). Over the decades various inventors have experimented with sets of contra-rotating rotors, which in their minds squeeze more energy out of the wind. Many years ago one German manufacturer even put such a product on the market. But Selsam places both his rotors on the same shaft so they rotate in the same direction. While he swears by them, their performance seems to be about what you would expect from two rotors of that size. Still,

it's hard to be any more novel than using twin rotors.

Selsam represents those determined inventors who have an idea and carry it through to completion, regardless of naysaying experts (the author included). Unlike most inventors, though, Selsam does his homework, tests his product in the field, and measures the results. Sure, he also drives his truck down Southern California roads with his turbine mounted on top, to howls of derision from wind professionals (and probably the California Highway Patrol as well). But Selsam tests his turbine in the wind on

Figure 1-24. Vertica Salad Spinner. This "novel" vertical-axis wind turbine is installed on a rooftop within Montreal's Biosphere. The manufacturer says the turbine is capable of 3 kW, about three times what can be reasonably expected.

top of a tower, just as if installed by a consumer. And unlike other inventors—and many small turbine manufacturers as well—he doesn't hide the results from the public. That's gutsy for a newcomer with an unproven technology.

Ventilators and Squirrels in a Cage

Many "novel" wind turbines are nothing more than rooftop ventilators repackaged as "wind turbines" or—as Mick Sagrillo sarcastically calls them—"salad spinners on a roof" (see figure 1-24, Vertica "salad spinner"). As ventilators, they work fine. It's when you try to couple them to a generator that you quickly learn why wind turbines use two or three slender, airfoil-shaped blades. Most inventors, however, never progress that far. They rarely build actual wind turbines, and perchance they do build one, they never measure its performance. Of course, they wildly exaggerate the potential of these breath-taking new inventions.

Another perennial favorite is, for lack of a better term, the squirrel-cage rotor. This uses an outer ring of vanes that are intended to direct the flow over an inner vertical-axis rotor, like a Francis hydro turbine (see figure 1-25, Squirrels in a cage). As with many other "new" wind turbine inventions, squirrel-cage rotors never work as well as advertised.

Doug Selsam, himself an inventor, has tried to understand why consumers—and the news media—are so gullible. His explanation: Ventilators and squirrel-cage rotors are easy to understand; modern wind turbines, much less so. After all, a rooftop ventilator looks like it will capture more wind than a modern wind turbine with only a few blades, or—unbelievably—only one.

In one Internet scam, a company peddling ventilators as "wind turbines" claimed that its product would produce nearly five times more electricity than a conventional wind turbine of the same size. Naturally, for this "superior" performance it would charge two to three times more than for a real wind turbine. The company asserted that it was "thinking outside the box," a catchphrase of 1990s management gurus. It certainly was. It was not even close to the box. It was on another planet where the laws of physics don't apply.

Figure 1-25. Squirrels in a Cage. This example of a squirrel-cage vertical-axis wind turbine has stood along the I-210 near San Fernando, California, since the mid-1980s. The outer ring of slats supposedly directs the flow over an inner rotor.

Ducted or Augmented Turbines

Unlike Selsam's novel twin rotor, ducted wind turbines were invented and reinvented throughout the 20th century. One such inventor was Dew Oliver, who installed his "blunderbuss," or ducted turbine, in California's San Gorgonio Pass in the 1920s. Historian Robert Righter notes that though Oliver claimed the turbine worked, that didn't prevent Oliver from being eventually convicted of fraud.

While many inventors opt for squirrel-cage rotors as a surefire means for extracting more energy from the wind than conventional wind turbines, others, such as Dew Oliver, opt for wind turbines encased in shrouds or ducts. As with squirrel-cage rotors and their curved deflectors, shrouded turbines are easy to visualize. Much like giant funnels, the shrouds concentrate or *augment* the flow across the wind turbine's rotor (see figure 1-26, Ducted turbine).

However, ducted turbines are often wrapped in mysterious technobabble, such as *diffuser augmentation,* that is beyond the ken of most observers. Even supposedly sophisticated engineers have been snared by what at first appears to be a startling new technology "overlooked" for decades by everyone else.

Do augmented turbines work? Yes, of course. Cup anemometers work, too, but we don't use them to produce electricity. Why? Because cup anemometers can't compete with modern, high-speed turbines. The same is true with ducted turbines. They have been tried—time and again—and found wanting.

Diffuser-augmented turbines can achieve conversion efficiencies much higher than conventional wind turbines with rotors of the

Figure 1-26. Ducted Turbine. Ducted, or augmented, turbines always promise a revolutionary way to extract more energy from the wind than conventional wind turbines. This promise has yet to be fulfilled—despite many attempts.

same diameter. Consequently, ducted turbines periodically reappear in both the trade and professional press as a "promising" new technology—often in conjunction with a jump in the price of oil.

What is missed, however, is that a ducted turbine uses, well, a duct, and it is this big duct, or shroud, that increases the area intercepted by the ducted wind turbine relative to that of a conventional wind turbine of the same rotor diameter. And someone has to pay for that big shroud around the rotor, and pay to turn the whole assembly into the wind.

One outspoken critic of diffuser augmentation is Professor Heiner Dörner at the University of Stuttgart's Institute of Aircraft Design. Sure, Dörner says, wind tunnel tests show you that a duct can double the wind speed across the rotor. But for this to happen, the wind must flow directly into the concentrator, a condition found in a wind tunnel—but only rarely in the real world. Yes, he says, the complete ducted assembly can turn to face changes in wind direction, but rarely accurately enough. Thus, the concentrating effect will be difficult, if not impossible, to achieve in operation.

This is one of the results seen by the few ducted turbines that have ever been built. Engineers first noted in their reports that yes, they were able to augment the flow (boost efficiency) as promised in the sales brochures; they went on about 120 watts, not 500 watts. By this criterion, the turbine claims it can produce five times more power than other wind turbines of its size. Another way to look at this is the rotor loading at rated power claimed by the manufac-

---

Over the years augmented and ducted turbines have never produced the amount of energy promised at the cost promised.

---

to say, however, that for various reasons—there were always extenuating circumstances—the concentrating effect was much less than they had anticipated. Moreover, there were other problems as well. The shroud was more expensive and difficult to construct than they first thought, turning the ungainly turbine into the wind was more difficult, and so on.

In the fall of 2005 a Swiss company was displaying a "new" ducted turbine, the Enflo WindTec 0071, at the big wind extravaganza in Husum, Germany. The turbine is small and well suited for an indoor display. The rotor is only 0.71 meter in diameter. The shroud, which surrounds the rotor, is 0.87 meter in diameter. With the shroud included, the Enflo 0071 is about the size of an Ampair 100. For comparison, most turbines at the Husum trade show are so big, they build the tents around the nacelles.

However, unlike the Ampair 100, a conventional micro turbine rated at 100 watts, the Enflo 0071 is rated at 500 watts. Let's examine whether this claim is realistic or not.

First, we'll consider only the area of the wind intercepted by the shroud, about 0.6 m^2, not the smaller area of the rotor inside the shroud. If a turbine of this size performed like other turbines in this class, it would be capable of

turer. The rotor loading for this turbine is 840 W/m^2 of shroud intercept area compared with the typical 250 to 300 W/m^2.

Second, let's compare the manufacturer's claimed performance with the Betz limit. The manufacturer claims that the Enflo 0071 will convert 70 percent of the energy in the wind at its rated power. Albert Betz, a German engineer, argued mathematically that the maximum we can theoretically extract from the wind is 59 percent. Even with the shroud included, the Enflo 0071 exceeds the Betz limit.

How then does the Enflo ducted turbine compare with other turbines? Its claims, and these remain just claims, exceed by a wide margin the performance of other turbines in its size class.

Can the Enflo 0071 do what it claims? The manufacturer's claims are not so far from what's theoretically possible, it is conceivable. Is it likely? No. The odds are long that this turbine can come even close to delivering on its promise of outsize performance. The burden of proof is always on the manufacturer to demonstrate that its turbine can do what it claims, more so in a case like this where the claims are at the theoretical limits.

Over the years augmented and ducted

turbines have never produced the amount of energy promised at the cost promised. They have never fulfilled their often highly touted claims. Modern wind turbines, for all their towers raise the turbines above the treetops. The mature trees in the Black Forest could be 30 meters (100 ft) tall. The nearly 100-meter tall tower may raise the turbines 100 meters (328 ft)

"If you can't afford to put it on a tall tower, then it's not a serious wind generator." —Mick Sagrillo

---

limitations, reliably deliver quantities of electricity at increasingly competitive costs. In short, conventional wind turbines work, and they work better than the few ducted turbines that have been tested.

What lessons can you take away from this? As with most things in life, don't believe everything you read. There is a deep-seated desire in all of us to believe we can get something for nothing. In wind energy this trait manifests itself in our willingness to suspend critical judgment—that voice in us asking, *Does this really work, and if it does where are the results?* Even the technically trained sometimes fall for the latest fad and get swept up in the madness of crowds. In wind energy, it's *Caveat emptor.* Let the buyer beware.

Towers

Wind turbines need towers to place the rotor above nearby obstructions. As we've seen throughout this chapter wind speed, and hence power, increases dramatically with height above nearby obstructions. We often think of towers raising wind turbines above the ground, but it's more than the ground level that is of concern. It is the relative level of nearby obstructions. For example, in figure 7-6, the unusually tall

above the ground, but only 70 meters (230 ft) above the treetops. It is this 70-meter difference that matters. Even the VAWTs installed on rooftops have short stub towers to raise the turbine above the parapet.

"The fuel is up there, not down here," says the irascible Mick Sagrillo, who for decades has been urging tall towers for even the smallest of small wind turbines. "If you can't afford to put it on a tall tower, then it's a serious wind generator."

Small wind turbines are installed on freestanding lattice towers similar to those of traditional farm windmills, free-standing tubular towers, or guyed masts (see figure 1-27, Tower types for small wind turbines). Tubular towers may be either straight-walled or tapered. Guyed masts use either lattice tower sections, pipe, or tubing, depending upon the design.

Tapered tubular towers are the most visually pleasing, but also the most costly. Also, unless they're hinged for lowering to the ground, they're also the most difficult to climb for servicing the turbine (see figure 1-5, Downwind wind turbine). Freestanding lattice towers are nearly as expensive.

Guyed towers are least expensive. For decades Bergey Windpower has used guyed lattice masts for its household-size turbines. For minor

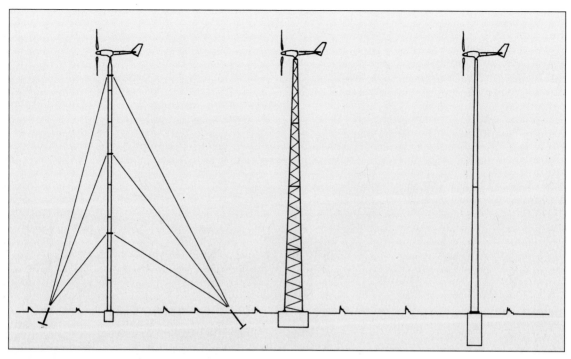

Figure 1-27. Tower Types for Small Wind Turbines. Small wind turbines are installed on guyed towers, freestanding lattice masts, and freestanding tubular towers. (Bergey Windpower, www.bergey.com)

repairs, you can climb the latticework to the top of the tower, rather than lowering the entire mast and turbine to the ground.

If they're hinged, guyed towers can be raised and lowered for installing and servicing the turbine. The boom in micro and mini turbines is partly due to the design of kits of inexpensive, guyed masts using steel pipe and thin-walled steel tubing that can be easily raised and lowered.

Micro wind turbines should nearly always be installed on tilt-up guyed towers. Their limited generation doesn't justify more elaborate towers. Many of the smaller mini wind turbines should likewise be installed only on hinged, guyed towers.

As turbines increase in size, more tower options are available. Household-size turbines are productive enough to justify freestanding towers, or heavier guyed towers. Small commercial-scale wind turbines are typically installed on freestanding lattice towers or tubular towers. Similarly, large wind turbines are installed on truss towers, tubular steel towers, or hollow concrete towers.

Wind Energy Basics

Let's start at the beginning, finding the power in the wind. Then we'll move on to calculating the area swept by a wind turbine rotor and the increase in wind speed with height above the ground—the reason wind turbines are installed atop tall towers.

Power in the Wind

The power (P) in the wind is a function of air density (ρ), the area intercepting the wind (A), and the cube of the instantaneous wind velocity (V^3), or speed.

$$P = \frac{1}{2}\,\rho A V^3$$

Increase air density, intercept area, or wind speed, and you increase the power available from the wind. Power is directly proportional to changes in air density and the area of the wind stream. Double the area, and you double the power. Slight changes in wind speed, however, significantly change the power available. Double the wind speed, and you increase the power available by eight times ($2^3 = 2 \times 2 \times 2 = 8$)!

If the value for air density in kg/m^3 at sea level is substituted for ρ in the equation, power in watts is

$$P = \frac{1}{2}(1.225)AV^3$$

$$P = 0.6125\ AV^3$$

Where speed is in meters per second (m/s) and area is in square meters (m^2).

Air Density

Air density varies with temperature and elevation. Warm air is less dense than cold air. Any given wind turbine will produce less in the heat of summer than it will in the dead of winter, in winds of the same speed. Minnesotans seeking wind development to offset an aging nuclear plant proudly boast that the upper Midwest's frigid winter winds hold more power than the equivalent winds in hot Southern California. Similarly, there is less power in the wind for a specific wind speed at a mountaintop telecommunications site in British Columbia than there is near sea level at the Folkecenter for Renewable Energy on Denmark's Skibsted Fjord.

Changes in air density relative to standard conditions at sea level can cut power production 10 to 20 percent, sometimes even more. For example, on a hot summer day atop 1,500-meter (5,000 ft) Cameron Ridge in the

Tehachapi Pass, the air temperature can reach 35°C (95°F). Under these conditions, the air is 80 percent as dense as along the coast near Santa Barbara at a comfortable 20°C (68°F). Thus, wind professionals must always take elevation and average temperatures into account when estimating how much electricity a wind turbine will produce. But the effect of changes in temperature or elevation on wind power can be dwarfed by changes in wind speed.

The Cube of Wind Speed

It bears repeating that the power in the wind varies with the cube of wind speed. Even a small increase in wind speed substantially boosts the power in the wind. This is the reason for all the fuss about siting wind turbines to take best advantage of the wind.

Consider the power available at one site with a wind speed of 4 (the units don't matter here) and another site with a wind speed of 5. The wind at the windier site is only 25 percent greater (5/4 = 1.25), yet there is 95 percent more power available—or nearly twice as much.

$$P_2/P_1 = (V_2/V_1)^3$$

$$P_2 = (5/4)^3 P_1 = 1.95\ P_1$$

Swept Area

As we've seen, power is directly related to the area (A) intercepting the wind. Double the capture area and you double the power available. Consider a conventional wind turbine, where the rotor spins about a horizontal axis. The rotor sweeps a disk, the area of a circle,

$$A = \pi r^2$$

where area (A) equals the product of π and the square of the rotor's radius (r).

This relationship between the rotor's radius (or its diameter) and energy capture is fundamental to understanding wind turbines, regardless of whether the rotor spins about a horizontal or vertical axis. Knowing this, you can quickly size up any wind machine by noting the dimensions of its rotor.

For conventional wind turbines, relatively small increases in blade length produce a correspondingly large increase in swept area, and thus in the power available.

Compare the area swept by one wind turbine with a rotor diameter of 10 and that of another with a rotor diameter of 14. (Again, the units are not important here.)

$$A_2 = (r_2/r_1)^2\ A_1$$

$$A_2 = (14/10)^2\ A_1 = 1.96\ A_1$$

Increasing the rotor diameter by 40 percent (from 10 to 14) increases the capture area by 96 percent—or nearly twice as much.

This exponential relationship between swept area and the power available also explains a crucial wind energy axiom: For conventional horizontal-axis wind turbines, nothing tells you more about a wind turbine's potential than rotor diameter—nothing. The wind turbine with the bigger rotor will almost invariably generate more electricity than a turbine with a smaller rotor, regardless of generator ratings or purported efficiency.

Whether we're talking about a conventional wind turbine or a vertical-axis wind turbine, it is the area swept by the rotor that matters. For

battery-charging systems, and for small wind turbines generally, average monthly wind speeds are also commonly used.

Because power is a cubic function of wind speed, periods of strong winds contribute far

---

The wind turbine with the bigger rotor will almost invariably generate more electricity than a turbine with a smaller rotor, regardless of generator ratings or purported efficiency.

---

a vertical-axis wind turbine that uses straight blades in an H-configuration, the area swept by the rotor is simply the height or length of the blades times the diameter. If you double the length of the blades, and you double the diameter, you increase the swept area by four (2 × 2 = 4).

In this book, wind turbines are ranked by swept area, not by generator capacity.

Wind Speed Distribution

Though we frequently speak in terms of *power* in referring to wind turbines, it's often just a substitute for *energy*. Energy is power over some unit of time. After all, it's energy that we're after. It's energy in kilowatt-hours (kWh) that we store in the batteries of an off-the-grid wind system, or energy in kWh that we sell to the utility for profit from a commercial wind turbine. Thus, we need to better define the wind speed used in the power equation to represent the speed of the wind over time.

We commonly designate the wind resource at a site by its average wind speed: typically, its average annual wind speed. In off-the-grid

more to annual energy production than would be indicated by the average wind speed alone. Thus it's critical to know the distribution of wind speeds throughout the period, whether for one month or for one year.

Jack Park, one of the pioneers in America's 1970s wind power revival, put it succinctly: "The average of the cubes is greater than the cube of the average." That is, the average of the cube of different wind speeds over time is greater than the cube of the average speed. To account for this, we need to know the actual distribution of wind speeds over time, assume a hypothetical distribution, or otherwise compensate with what Park has called the "cube factor" and others have labeled the "Energy Pattern Factor."

The power in the wind at three different sites with exactly the same average wind speed illustrates the importance of Park's "cube factor" (see table 2-1, Effect of Speed Distribution on Wind Power Density for Sites with Same Average Speed). Though the New York site experiences the same average wind speed as the one in Puerto Rico, the Caribbean island lies in the trade-wind belt and has more constant winds. These steady winds produce less power over time than a temperate wind regime like

Table 2-1: Effect of Speed Distribution on Wind Power Density for Sites with Same Average Speed				
	Annual Average Wind Speed		Wind Power Density	Energy Pattern Factor or	Battelle Wind Power Class
Site	m/s	mph	W/m²	Cube Factor	(at 10 m)
Culebra, Puerto Rico	6.3	14	220	1.4	4
Tiana Beach, New York	6.3	14	285	1.9	5
San Gorgonio, California	6.3	14	365	2.4	6
Battelle, PNL Wind Energy Resource Atlas, 1986.					

that of New York or the blustery winds that rush through California's San Gorgonio Pass. The winds of the San Gorgonio Pass contain 66 percent more power than the gentler winds bathing Puerto Rico. They are nearly 30 percent more powerful than winds at the stormy shoreline of Long Island.

Meteorologists have characterized the distribution of wind speeds for many of the world's wind regimes. For temperate climates such as that of continental North America, the Rayleigh wind speed distribution offers a good approximation. Like New York's Tiana Beach, the cube or energy pattern factor for the Rayleigh distribution is 1.9 (see figure 2-1, Rayleigh Wind Speed Distribution).

Wind Resources

There is just no escaping the fact that the amount of wind you have at your site determines how much power you can expect from a wind turbine of a given size. Though few would ever consider placing a solar panel in the shade and expecting it to generate electricity, it's surprising the number of people who try to install a wind turbine sheltered from the wind—often by using too short a tower.

Because small changes in wind speed have such a profound effect on the power in the wind, many in the past have recommended measuring wind speeds for a minimum of one year prior to deciding whether or not to install

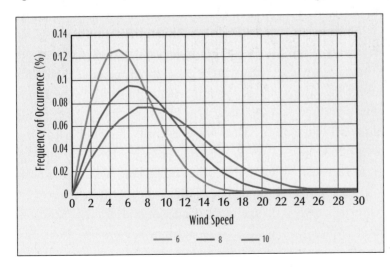

Figure 2-1. Rayleigh Wind Speed Distribution For Average Annual Wind Speed. This chart shows the occurrence of winds at different average speeds. The units are not important here. Charts for wind speeds in mph and m/s are similar. For example, at an average annual wind speed of 6, winds of a speed 4 occur 12 percent of the time or 1,050 hours per year.

a wind turbine. Commercial wind power developers in many parts of the world do just that. They install an anemometer mast and monitor the winds with an electronic data logger. Then they compare their results with long-term records from nearby sites (see figure 2-2, Sketch of guyed anemometer mast and instruments; and figure 2-3, Wind data loggers).

This is impractical for people who want to use a battery-charging wind turbine off the grid, simply because the cost of the mast, measuring devices, and necessary professional analysis will often exceed the cost of a small wind turbine.

For years Mick Sagrillo, a small wind turbine guru in Wisconsin, has suggested that it would be better to simply install a micro turbine and monitor its performance. If you're dissatisfied with its production, just take it down and sell it. There's a ready market for used wind turbines in the classified ads of *Home Power* magazine, but less of a market for used anemometers.

Many purchasers of micro turbines have followed this advice and opted for the small machines over measuring the wind with a recording anemometer. Their reasoning is simple. *My*

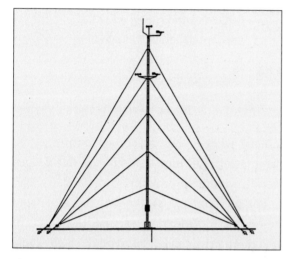

Figure 2-2. Guyed Anemometer Mast and Instruments. (NRG Systems, www.nrgsystems.com)

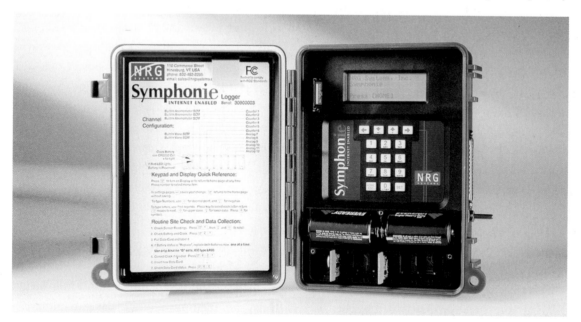

Figure 2-3. Wind Data Logger. Electronic data recorders can be used to collect wind data or, with the appropriate sensors, to monitor wind turbine performance. (NRG Systems, www.nrgsystems.com)

photovoltaic panels don't give me enough power in the winter and I need a supplemental power source. A small wind turbine could be helpful, and I can buy a micro turbine for about what I'd pay for a recording anemometer. In the meantime, I'll get something usable, kilowatt-hours.

Like Sagrillo in Wisconsin, Vision Quest Windelectric's Jason Edworthy says that small wind turbines seldom justify full-blown wind resource assessments, especially for off-the-grid use. In most stand-alone applications, the high value that a wind turbine adds to a hybrid wind and solar system warrants its use, often regardless of the wind resource. There are exceptions, of course. It makes no sense, for example, to install a wind turbine in a forest where trees will block the wind, just as it makes no sense to mount a solar panel under the shade of an awning.

Another approach to determining your wind resource is to locate owners of small wind turbines in your vicinity and determine how well their machines have performed.

Nevertheless, using a small wind turbine as an anemometer to measure the wind is at best unreliable, says Detronic's Svend de Bruyn. He tested the approach at his site north of Toronto, Canada. What he found was that the performance of a micro turbine was a very poor indicator of actual wind speeds. While operating a small wind turbine may give an indication of how much electricity you can expect from a bigger turbine, it will not tell you the average wind speed or exactly how much more you can expect.

This doesn't mean you shouldn't study the wind at your site. For distributed applications that use commercial-scale turbines, the investment is so great that it behooves you to know

North American Wind Resource Maps

- US DOE's Wind Powering America Wind Resource Maps:
  www.eere.energy.gov/windandhydro/windpoweringamerica/wind_maps.asp
- Canada's Wind Energy Atlas:
  www.windatlas.ca/en/maps.php
- Wind Resource Maps of Mexico:
  www.re.sandia.gov/en/ti/rm/rm-rt.htm

exactly what to expect. For medium-size and large wind turbines especially, a full-fledged wind-monitoring program is required.

For preliminary assessments, maps of wind resources for most states and provinces in North America can be found on the Web. These can help you determine whether there is sufficient justification for further exploring the use of wind energy, or for spending the money on a full wind-measuring program at a specific site.

Wind resource maps, or atlases, as they're sometimes called, use wind power "classes" to designate areas of different wind potential. These classes are surrogates for a range of wind power densities, or wind power per square meter of the wind stream (W/m^2) at a particular height above the ground. For example, a region with a class 4 resource at 30 meters (100 ft) above the ground has a wind power density from 320 to 400 W/m^2, or the equivalent of annual wind speeds from 6.5 to 7.0 m/s (15–16 mph). Make no mistake, class 4 is very windy. At 50 meters (160 ft), the minimum tower height today for medium-size and large commercial wind turbines, class 4 sites have a wind power density of 400 to 500 W/m^2, or the equivalent of

Table 2-2: Battelle Classes of Wind Power Density								
	Wind Speed and Power at 10 m (33 ft)			Wind Speed and Power at 30 m (100 ft)			Wind Speed and Power at 50 m (164 ft)		
	Power Density	Speed		Power Density	Speed		Power Density	Speed	
Wind Power Class	W/m²	m/s	mph	W/m²	m/s	mph	W/m²	m/s	mph
	50	3.5	7.8	80	4.1	9.2	100	4.4	9.9
1									
	100	4.4	9.9	160	5.1	11.4	200	5.5	12.3
2									
	150	5.0	11.5	240	5.9	13.2	300	6.3	14.1
3									
	200	5.5	12.3	320	6.5	14.6	400	7.0	15.7
4									
	250	6.0	13.4	400	7.0	15.7	500	7.5	16.8
5									
	300	6.3	14.1	480	7.4	16.6	600	8.0	17.9
6									
	400	7.0	15.7	640	8.2	18.4	800	8.8	19.7
7									
	1000	9.5	21.3	1600	11.1	24.7	2000	11.9	26.7

Note: Increase in speed and power with height assumes 1/7 power law. Wind speeds are based on the equivalent of a Rayleigh distribution.

7 to 7.5 m/s (16–17 mph). See table 2-2, Battelle Classes of Wind Power Density.

Increases in Wind Speed and Power with Height

As indicated in the previous discussion of wind resource classes, the height above the ground greatly influences the amount of wind available. Because obstructions near the ground disrupt the flow of the wind, wind speeds increase with height. Over rough terrain, wind speeds can sometimes increase dramatically. This effect is so important that data on wind speeds will always include the height at which the wind was measured. If the height is not specifically mentioned, it is usually assumed to be about 10 meters (33 ft) above the ground. Most wind turbines will be installed on towers much taller than this to take advantage of the stronger, less turbulent winds aloft.

The easiest way to calculate the increase in wind speed with height is to use the "power law" method. The increase in wind speed with height can be found from

$$V/V_o = (H/H_o)^\alpha$$

$$V = (H/H_o)^\alpha V_o$$

where V_o is the wind speed at the original height,

V is the wind speed at the new height, H_0 is the original height, H is the new height, and α is the surface roughness exponent.

The rate at which wind speeds increase with height varies with surface roughness: the sum of the effects of vegetation and terrain. The increase is greatest over rough terrain or numerous obstacles, such as trees and shrubs, and smallest over smooth terrain, such as the surface of a lake. In the power law equation, this is reflected in the roughness exponent, α (see table 2-3, Surface Roughness and the Wind Shear Exponent).

Over smooth terrain α may be less than 0.1, and over rough terrain as great as 0.4. On much of the Great Plains, the steppes of North America, the 1/7 power law often applies. That is, the roughness exponent is 1/7 or 0.14.

Consider the increase in wind speed when doubling tower height from 10 to 20 or from 30 to 60. The units are unimportant. It's the ratio that counts.

$$V = (20/10)^{0.14}\,V_0 = 2^{0.14}\,V_0 = 1.1\,V_0$$

On terrain where the 1/7 power law applies, doubling tower height increases wind speed by 10 percent. Increasing tower height five times, say from 10 to 50 or from 30 to 150, may increase wind speed as much as 25 percent (see figure 2-4, Increase in Wind Speed with Height).

$$V = (50/10)^{0.14}\,V_0 = 1.25\,V_0$$

But power increases even more dramatically because of its cubic relationship with speed. Doubling tower height increases the power available by 34 percent.

$$P = (H/H_0)^{3\alpha}\,P_0 = (2)^{3(0.14)}\,P_0 = 1.34\,P_0$$

So far, so good. Now, watch what happens to

Table 2-3: Surface Roughness and the Wind Shear Exponent α	
Terrain	Wind Shear Exponent α
Ice	0.07
Snow on flat ground	0.09
Calm sea	0.09
Coast with on-shore winds	0.11
Snow-covered crop stubble	0.12
Cut grass	0.14
Short-grass prairie	0.16
Crops, tall-grass prairie	0.19
Hedges	0.21
Scattered trees & hedges	0.24
Trees, hedges, a few buildings	0.29
Suburbs	0.31
Woodlands	0.43

Relative to a reference height of 10 m (33 ft)

Adapted from Characteristics of the Wind by Walter Frost and Carl Aspliden in *Wind Turbine Technology*, and *Windenergie: Theorie, Anwendung, Messung* by Jens-Peter Molly.

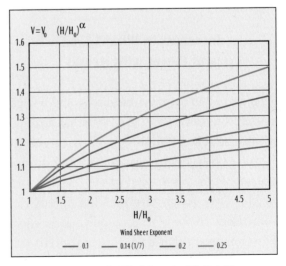

Figure 2-4. Increase in Wind Speed with Height.

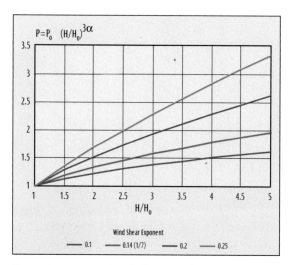

$$P = P_0 \ (H/H_0)^{3\alpha}$$

Wind Shear Exponent
— 0.1 — 0.14 (1/7) — 0.2 — 0.25

Figure 2-5. Increase in Wind Power with Height.

Table 2-4: Increase in Wind Speed and Power with Height		
$\alpha = 1/7$ (0.14) for low grass		
	2 x Height	5 x Height
Wind Speed	1.1	1.25
Wind Power	1.35	1.99

the power in the wind when we increase tower height five times. The power available nearly doubles (see figure 2-5, Increase in Wind Power with Height).

$$P = (5)^{3(0.14)}P_0 = 1.97 \ P_0$$

This is why wind turbines are installed on tall towers. If you want a productive wind turbine, use as tall a tower as you can afford (see table 2-4, Increase in Wind Speed and Power with Height).

It's worth repeating. Doubling the height of the tower often increases the power available by as much as 25 percent. Increasing the height of the tower five times doubles the amount of power available. Taking shortcuts by using short towers—or worse, mounting the wind turbine on the roof—nearly always leads to disappointment.

In the next chapter we use what we've learned here to calculate how much energy wind turbines can theoretically capture.

Estimating Performance

Be advised. While estimating how much energy a wind turbine might produce at a given site involves some minor number crunching, the results are not exact. This is not rocket science. Small wind turbines in particular are notorious for defying expectations, especially in battery-charging systems. Even the experts have trouble.

Large wind turbines linked to the utility grid feed an infinite sink. The utility system will consume all the energy the turbine produces. When the wind turbine captures the wind's energy and converts it to electricity, the turbine can deliver all of its generation to the grid. Thus the electrical load on the wind turbine is predictable.

The situation with small battery-charging wind turbines is just the opposite. The load—the batteries—may not always be able to take the energy when it's available. When the batteries become fully charged, the turbine must spill or dump the excess energy that's available. Some manufacturers of micro and mini wind turbines offer dump loads—circuits for using this excess generation and keeping the wind generator fully loaded. Often small wind turbines simply spill the energy available in the wind when the batteries become fully charged. Thus it's difficult to estimate the performance a consumer can expect from a small wind turbine, because the turbine may be spilling the energy in the wind instead of delivering it to a load.

Another factor is that most small wind turbine manufacturers lack the funds necessary to perform extended field tests on their products. These are small businesses with small staffs that have little time or money for the painstaking tests needed to understand how their products perform in the field.

With these caveats in mind, we'll examine three methods for calculating the gross amount of energy that wind turbines—of any size—may capture. The first uses the swept area of the rotor. The second uses a curve of the power a wind turbine will produce at various wind speeds. The third approach simply uses the manufacturer's published estimates.

Swept Area Method

This is a back-of-the-envelope technique that wind professionals use to quickly size up a wind turbine—and often to size up the people promoting it. If a salesman says any wind turbine will deliver two or three times more energy than that calculated with this method, show him the door.

Previously we learned that rotor diameter, or more correctly the rotor's swept area, is one of the critical factors in determining how much

Debunking Pyramidal Power and Magical Mag-Wind

The pyramids have always been thought to contain some magical power. Maybe they do. Maybe their magic can be applied to wind turbines. Then again, maybe not.

In the mid-2000s the news media was abuzz—not just the Internet, where new "revolutionary" inventions are a plague, but also mainstream sources that should know better—about a new "magnetically levitated vertical-axis wind turbine," the Mag-Wind.

The "invention" had all the telltale signs: hype high, experience low (actually nonexistent). Other tip-offs were claims that the device was "much smaller than other wind turbines," and that it was "much more efficient than the old-fashioned windmill."

Let's look at some of Mag-Wind's claims and see how these can be manipulated.

Mag-Wind might indeed be more efficient than a "old-fashioned windmill," but will it be more

Figure 3-sb1. The Mag-Wind on a rooftop in Grimsby, Ontario. There was a good stiff wind at the time this photo was taken on July 12, 2007, and the turbine was not spinning.

efficient than a modern, highly efficient wind turbine? That's very doubtful. Modern wind turbines look the way they do for a good reason. They squeeze a lot of energy out of the wind with very little materials—those long slender blades.

Next, we'll check Mag-Wind's numbers, first by using the swept area method. To begin, let's simply forget Mag-Wind's conical shape and assume that it's your garden-variety vertical-axis wind turbine that uses an H-rotor. This is to Mag-Wind's advantage—and it needs every advantage to meet its claims.

Mag-Wind asserted that its turbine could generate 13,200 kWh per year in an average wind speed of 13 mph (5.8 m/s). First off, that seems like an awful lot for a windmill that will be sitting on your roof, as shown in their ads.

The windmill was 6 feet (1.8 m) tall by 4 feet (1.2 m) in diameter at its widest point. Thus the rotor sweeps a paltry 2.2 m² (1.8 × 1.2 = 2.2).

At an average annual wind speed of 6 m/s (13.4 mph), typical small wind turbines will generate no more than 400 kWh/m²/yr. In Mag-Wind's case, then, it could produce a maximum 1,000 kWh per year (400 × 2.2 = 880). Ouch, Mag-Wind claimed that its revolutionary new windmill would produce 13 times the typical small turbine.

Okay, that may be too hard on the company. Modern large wind turbines can produce 770 kWh/m²/yr at average annual wind speeds of 6 m/s. So let's assume that despite its small size, the Mag-Wind performs at least as well as giant wind turbines. Mag-Wind's turbine could then produce some 1,700 kWh per year. Let's give the firm the benefit of the doubt, again, and round that up to 2,000 kWh per year. Therefore, Mag-Wind claimed that its turbine could produce more than six times that of the typical commercial-scale wind turbine.

And before someone begins to say, *Well, it could be more efficient than any wind turbine ever built,* let's try one more approach.

Mag-Wind says its wind turbine will produce 5 kW at a rated speed of 28 mph (12.5 m/s). The Mag-Wind is about the size of a conventional wind turbine with a rotor 1.7 meters (6 ft) in diameter. This size turbine would have a standard power rating of only 450 watts. For Mag-Wind to produce 5 kW at that wind speed, it would need to be nearly 200 percent efficient at doing so. That is, it must produce two times the amount of energy in the wind.

That's some windmill. It produces more energy than there is in the wind striking it!

To summarize, will it do what it claims? No, not on this planet. What is it good for then? A roof vent or a lawn ornament. It could make a nice lawn ornament. It's not a wind turbine.

Why is all this important? Because the media and political attention that is directed toward such "revolutionary inventions" distracts us from the real work at hand by offering us an easy way out of our energy problems. If only we'll use this new device, all will be well and we won't have to make any of the painful—and expensive—decisions that are ultimately necessary to meet our energy needs in a sustainable and environmentally acceptable way.

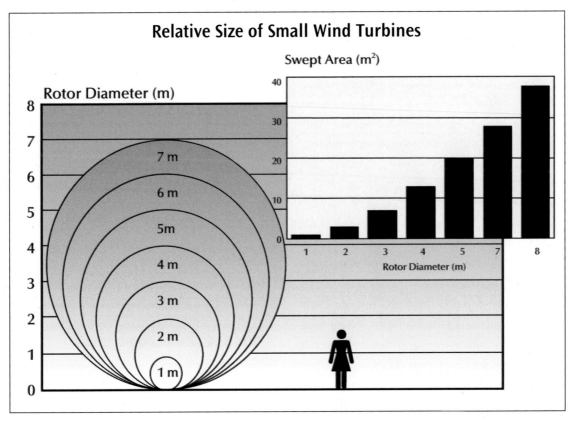

Figure 3-1. Relative Size of Small Wind Turbines.

energy a wind turbine can capture. The other critical factor is wind speed. If we know how typical wind turbines work under ideal conditions, we can use rotor swept area and average annual wind speed to estimate the annual energy output or annual energy production (see figure 3-1, Relative Size of Small Wind Turbines).

Let's assume we want to install Southwest Windpower's Air Breeze—the latest in the Air series of micro turbines—in the Texas Panhandle, where the average wind speed at hub height is 5.5 m/s or 12.3 mph (see figure 3-2, Air Breeze).

This wind turbine, among the smallest on the market, uses a rotor 1.2 meters (3.8 ft) in diameter and thus intercepts

Figure 3-2. The Air Breeze—a quiet successor to Southwest Windpower's popular Air series of micro turbines—at Wulf Test Field in California's Tehachapi Pass.

$$A = \pi r^2 = \pi(1.2/2)^2 = \pi(0.6)^2 = \pi(0.34) = 1.1m^2 \text{ of the wind stream.}$$

With this information and the formula for the power in the wind, we can calculate how much energy the wind turbine will intercept. But we don't know yet how much of that wind energy the turbine can capture.

Theoretically, the maximum performance a wind turbine rotor can achieve is 59 percent. This is the so-called Betz limit. Aerodynamically sophisticated rotors on large wind turbines can capture 40 percent and sometimes more of the energy in the wind. But this is just the rotor. There's more to a windmill than just the rotor.

Generators, especially those on small wind turbines, seldom produce more than 90 percent of the energy delivered to them. Throw in additional losses to compensate for the turbine yawing in response to changes in wind direction, losses in conductors carrying electricity down the tower, and so on, and you reach an overall conversion efficiency of about one-third, or 33 percent. See table 3-1, Estimated Annual Energy Production for Large Wind Turbines.

For example, the US Department of Agriculture (USDA) tested a Bergey 1500 for five years in a wind-electric pumping application at its Bushland experiment station west of Amarillo, Texas. Researchers found that the turbine was available for operation—the wind industry's measure of reliability—93 percent of the time. This is quite good for a small wind turbine, though there are examples where small turbines have operated trouble-free for several years. The Bergey 1500 in this application converted 23 percent of the energy in the wind into electricity in winds from about 8 m/s

Table 3-1: Estimated Annual Energy Production for Large Wind Turbines

Average Annual Wind Speed		Power Density	Total Efficiency	Average Annual Specific Yield
m/s	~mph	W/m²	η	kWh/m²/yr
4.0	9.0	75	0.35	230
4.5	10.1	107	0.36	340
5.0	11.2	146	0.37	470
5.5	12.3	195	0.36	610
6.0	13.4	253	0.35	770
6.5	14.6	321	0.33	930
7.0	15.7	401	0.31	1,090
7.5	16.8	494	0.29	1,230
8	17.9	599	0.26	1,360
8.5	19.0	718	0.24	1,480
9	20.2	853	0.21	1,570

Note: Gross generation for a single turbine at hub-height wind speed, based on published manufacturer's data. Actual performance will vary.

(18 mph) to 11 m/s (25 mph). The annual efficiency over the turbine's entire operating range was much less.

Typical small turbines will capture 20 percent or less of the annual energy available in the wind. There are several reasons for this. One is that the airfoils on small wind turbines are inherently less efficient than those on large wind turbines. The second may be due to the rapid yawing, or changing of direction, of small wind turbines in gusty winds.

Jim Tangler, an aerodynamicist who worked with wind turbines much of his professional life, has observed that small wind turbines are dynamically unstable in turbulent winds, like the winds found around homes and farm buildings. Because of their free yaw, "Many small wind turbines act just like large anemometers," says Tangler. They are always hunting the wind,

While vertical-axis wind turbines don't need to yaw to point into the wind, their performance to date hasn't shown that they are any more efficient at capturing the energy in the wind than conventional wind turbines.

first pointing this way, then that. In contrast, large wind turbines use mechanical drives to yaw them into the wind. Tangler, like others over the years, suggests that household-size or small commercial-size machines should use some form of damping to stabilize rapid yawing in turbulent winds.

While vertical-axis wind turbines don't need to yaw to point into the wind, their performance to date hasn't shown that they are any more efficient at capturing the energy in the wind than conventional wind turbines.

For whatever reason, micro, mini, and household-size wind turbines are less productive than commercial-scale turbines (see table 3-2, Estimated Annual Energy Production for Small Wind Turbines).

Now let's see what we can expect from Southwest Windpower's Air Breeze, a micro turbine, at a site with an average wind speed of 5.5 m/s (12.3 mph). The Air Breeze intercepts about 1.1 m² of the wind stream.

The annual energy (E) in the wind stream is

$$E = Pt = 1/2 \; \rho AV^3 \times 8,760$$

The annual energy output or annual energy production can be found using the equation for power in the wind for some period of time (t), a measure of the distribution of wind speeds during that time (1.9 for the Rayleigh distribution's energy pattern factor), and the conversion efficiency (η),[3]

$$AEO \; or \; AEP = 1/2 \; \rho AV^3 \times 8,760 \times 1.9 \times 20\%$$

in watt-hours per year. To find kilowatt-hours per year, we must divide by 1,000 W/kW.

Small wind turbines are typically designed to perform best in the low-wind regimes where most people live—at sites with average wind speeds of 4 to 5 m/s (9–11 mph), for example. In locales with higher average annual wind speeds, their performance drops off. It's not uncommon at extremely windy sites for small wind turbines to convert less than 15 percent of the energy in the wind. This is normal. Because of the cubic relationship with wind speed, there's so much energy available at windy sites that designers can afford to capture only a small part of it.

$$AEO \; or \; AEP = (½ \times 1.225 \times 1.1 \times 5.5^3 \times 8,760 \times 1.9 \times 20\%)/(1,000 \; W/kW) \sim 375 \; kWh/yr$$

This calculation is the principle behind both table 3-1 and table 3-2. The system efficiency in these tables is an average for each annual wind speed gleaned from the product literature of hundreds of different wind turbines. In table 3-2, for example, the average system efficiency of small turbines is 19

---

3. The conversion efficiency is sometimes written as C_p—but technically this refers to the aerodynamic efficiency of the rotor. We are interested here in the overall conversion efficiency of the entire wind turbine system, η.

Table 3-2: Estimated Annual Energy Production for Small Wind Turbines				
Average Annual Wind Speed		Power Density	Total Efficiency	Average Annual Specific Yield
m/s	~mph	W/m²	0	kWh/m²/yr
4.0	9.0	75	0.190	120
4.5	10.1	107	0.195	180
5.0	11.2	146	0.200	260
5.5	12.3	195	0.190	320
6.0	13.4	253	0.180	400
6.5	14.6	321	0.175	490
7.0	15.7	401	0.170	600
7.5	16.8	494	0.160	690
8	17.9	599	0.150	790
8.5	19.0	718	0.145	910
9	20.2	853	0.140	1,050

Note: Gross generation for a single turbine at hub-height wind speed, based on published manufacturer's data. Actual performance will vary.

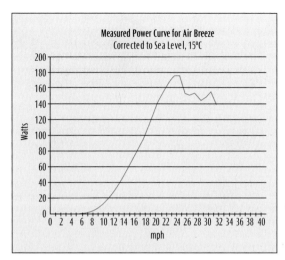

Figure 3-3. **Measured Power Curve for Air Breeze.** This power curve is derived from measurements made at the Wulf Test Field in the Tehachapi Pass in early 2008. The data was corrected to sea level standard conditions. Thus, this is not the manufacturer's data, but data measured by an independent third party. Results from similar tests by Appalachian State University are comparable. Power is shown on the vertical axis, and wind speed on the horizontal axis. Note that the power shown for each wind speed is averaged over a period of time from many different measurements. Measurements for commercial wind turbines use 10-minute periods; for small turbines, they're averaged over 1-minute intervals.

percent. Some wind turbines, such as the Air Breeze, will perform better than the average; others, less well.

In the following calculations, we'll find that the Air Breeze performs slightly better than the average based on actual field measurements in California's Tehachapi Pass.

Power Curve Method

The second method for estimating annual performance is the same as that used by the manufacturers themselves when calculating what their products will do under various wind conditions. They use a curve, or more correctly, a table of the power produced at different wind speeds (see figure 3-4).

Again, a word of warning. The power curves proffered by some manufacturers of small turbines can best be characterized as informed guesswork. View power curves and the energy calculations that result with a good dose of skepticism. There are no government agencies ensuring the accuracy of published power curves, though there should be. Wherever

Wherever possible, use power curves that have been verified by independent third parties, preferably professional testing laboratories or universities.

Table 3-3: Annual Energy Production for Air Breeze from Power Curve Measured at Wulf Test Field, Rayleigh Frequency Distribution

	Rotor Diameter		Area		Average Wind Speed	
	m	ft	m2		m/s	mph
	1.17	3.83	1.07		5.4	12

Wind Speed Bin mph	Measured Power at 15 C Sea Level W	Raleigh Frequency of Occurrence	Hours/Year	Energy kWh/yr	Gross Energy kWh/yr	System Efficiency
0	0.0	0.000	0	0	0	n/a
1	0.0	0.011	95	0	0	0.00
2	0.0	0.021	187	0	0	0.00
3	0.0	0.031	273	0	0	0.00
4	0.1	0.040	350	0	1	0.03
5	0.1	0.048	417	0	3	0.02
6	0.5	0.054	471	0	6	0.04
7	1.8	0.058	512	1	10	0.09
8	5.0	0.062	539	3	16	0.17
9	9.6	0.063	553	5	24	0.23
10	15.7	0.063	554	9	33	0.27
11	23.2	0.062	543	13	42	0.30
12	32.7	0.060	523	17	53	0.32
13	44.0	0.056	494	22	64	0.34
14	54.7	0.052	459	25	74	0.34
15	67.8	0.048	420	28	83	0.34
16	81.9	0.043	378	31	91	0.34
17	95.0	0.038	336	32	97	0.33
18	112.4	0.034	294	33	101	0.33
19	127.7	0.029	253	32	102	0.32
20	144.8	0.025	216	31	101	0.31
21	155.7	0.021	181	28	98	0.29
22	167.1	0.017	150	25	94	0.27
23	174.6	0.014	123	21	88	0.24
24	175.3	0.011	99	17	80	0.22
25	152.4	0.009	79	12	72	0.17
26	150.8	0.007	62	9	64	0.15
27	153.5	0.006	48	7	56	0.13
28	142.9	0.004	37	5	48	0.11
29	147.4	0.003	28	4	40	0.10

Table 3-3: Annual Energy Production for Air Breeze from Power Curve Measured at Wulf Test Field, Rayleigh Frequency Distribution (continued)						
Wind Speed Bin mph	Measured Power at 15 C Sea Level W	Raleigh Frequency of Occurrence	Hours/Year	Energy kWh/yr	Gross Energy kWh/yr	System Efficiency
30	155.5	0.002	21	3	34	0.10
31	138.9	0.002	16	2	27	0.08
32	0.0	0.001	11	0	22	0.00
33	0.0	0.001	8	0	18	0.00
34	0.0	0.001	6	0	14	0.00
35	0.0	0.000	4	0	11	0.00
36	0.0	0.000	3	0	8	0.00
37	0.0	0.000	2	0	6	0.00
38	0.0	0.000	1	0	4	0.00
39	0.0	0.000	1	0	3	0.00
40	0.0	0.000	1	0	2	0.00
Total		0	8,751	418	1,692	0.25

Note: Power curve corrected to sea level. Accuracy +/- 5%. For more on the measurements at the Wulf Test Field, see www.wind-works.org/articles/small_turbines.html.

possible, use power curves that have been verified by independent third parties, preferably professional testing laboratories or universities.

The values shown in the power curve are typically for the turbine's output at hub height.

The power curve will show power increasing with wind speed until a peak is reached. Subsequently, depending upon the design of the wind turbine, the power may remain fairly constant in stronger and stronger winds until a limit is reached, when controls are activated to reduce or spill power. At a given speed, the wind turbine begins governing. In the case of many small wind turbines, the rotor begins furling, or turning out of the wind. The drop in power can sometimes be dramatic, as the controls respond to protect the wind turbine from damage.

Thus the power curve will clearly show peak power. The manufacturer will specify the "rated power" as well. This is the wind speed at which the wind turbine produces the power on its nameplate. Rated power may not be apparent on the power curve.

Southwest Windpower rates the Air Breeze, the micro turbine in our example, at 200 watts at a wind speed of 12.5 m/s (28 mph). There are no standards for how or at what speed a manufacturer "rates" its wind turbine. This has led to untold confusion among consumers and professionals alike—for decades.

Never use "rated power" as a measure of performance or for estimating how much electricity a wind turbine will generate. Rated power is industry shorthand for the size of the wind turbine, and it's a lousy shorthand at that. For example, the Air Breeze is "rated" at 200 watts. Under field conditions, it doesn't produce its "rated power." At 28 mph in figure 3-3, Measured power curve for Air Breeze, the turbine produces about 150 watts.

While most small turbines fail to meet their "rated power," most large commercial-scale turbines do. When you spend $5 million for a wind turbine, you also have the money to ensure that the manufacturer honors its promises about performance.

Regardless of whether a wind turbine meets its advertised "rated power" or not, we are primarily interested in how much electricity it will produce. If we have a field-verified power curve, we can do that.

If we know the distribution of wind speeds over time, we can match the speed distribution with the power curve and then sum the energy produced by the turbine across the range of wind speeds represented as in table 3-3, Annual Energy Production for Air Breeze.

Measurements of the performance of Southwest Windpower's Air Breeze were conducted at the Wulf Test Field in California's Tehachapi Pass in early 2008. While the tests were not made by a professional laboratory, they conformed to much of the testing protocol used internationally. The results indicated that the peak power of the Air Breeze was closer to the manufacturer's rated power, 175 watts of its rated 200 watts, than any previous model. Most importantly, the measured power curve indicates that the Air Breeze could reasonably be expected to generate about 95 percent of the electricity projected by the manufacturer for average wind speeds of 5.4 m/s (12 mph). Tests at Appalachian State University's Beech Mountain site near Boone, North Carolina, found similar results.

Table 3-3 also includes the calculation of the gross energy in the wind at each wind speed. From this and the energy the turbine is expected to produce at each wind speed, we can calculate the wind turbine efficiency. In this example, the average efficiency of the wind turbine across the full range of wind speeds is 25 percent. Thus, the Air Breeze's actual measured performance is somewhat better than the average efficiency of 20 percent used in the swept area example.

Manufacturers' Estimates

If this is more math than you want to deal with, most manufacturers provide estimates of what to expect from their turbines under standard conditions, usually a Rayleigh distribution at sea level.[4] The format varies. Some manufacturers provide a chart; others provide a table. See table 3-4, Annual Energy Production at Hub Height.

Manufacturers of small wind turbines often provide estimates of average monthly generation. Southwest Windpower provides a chart of monthly generation. The values in table 3-4 are extracted from the chart and are approximations of what Southwest Windpower estimates the Air Breeze will deliver. Based on tests at the Wulf Field, the chart is accurate up to annual wind speeds of about 11 mph (4.9 m/s). At average wind speeds of 14 mph (6.2 m/s), the chart is off about 10 percent, which, for small wind turbines, is practically on the money.

Manufacturers of large wind turbines always provide estimates of AEO or AEP as annual averages.

---

4. Presented as a Weibull distribution with a k factor of 2.0.

Table 3-4: Southwest Windpower Air Breeze				
Wind Speed		AEO		
mph	m/s	~kWh/yr (measured)	% efficiency	kWh/yr (advertised)
9	4.0	207	0.29	204
10	4.5	277	0.28	264
11	4.9	349	0.27	348
12	5.4	418	0.25	444
13	5.8	482	0.23	516
14	6.2	537	0.20	600

Specifications: http://airbreeze.com/files/3-CMLT-1095_Air_Breeze_spec.pdf.

Now, for some more fun with numbers, let's take the performance Southwest Windpower projects for its Skystream 3.7 and see how it stacks up to the actual performance of its Air Breeze.

The Skystream uses a downwind rotor 3.72 meters (12.2 ft) in diameter.

$$A = \pi r^2 = \pi(3.72/2)^2 = \pi(1.86)^2 = \pi(3.5) \sim 10.9 \text{ m}^2$$

Thus it sweeps ~10.9 m² of the wind stream.

The wind power density at an average annual wind speed of 5.5 m/s is 195 W/m², and the specific energy in the wind for an entire year is 1,700 kWh/m² (195 x 8,760/1,000 = 1,700).

For a wind turbine that sweeps 10.9 m², there is 18,500 kWh/yr of gross energy available in the wind (10.9 x 1,700 = 18,500).

Southwest Windpower estimates that the Skystream will generate 4,800 kWh/yr at this average annual wind speed. Comparing its estimate (4,800) to the gross energy in the wind (18,500), the wind turbine—if it performs as the manufacturer suggests—converts 26 percent of the energy in the wind into electricity (4,800/18,500 ~ 26%). This is comparable to the measured performance of the Air Breeze at the Wulf Test Field, and is 7 percent greater than the 19 percent suggested in table 3-2.

No matter what technique is used, these projections represent the gross amount a turbine can be expected to produce. For a host of reasons, all of this energy is rarely put to use in a battery-charging system.

In battery-charging systems, batteries that are fully charged may not be able to take more energy when, in a good stiff wind, the turbine is churning it out. Further, some of the energy that is eventually stored in the batteries is lost due to the inherent inefficiency of battery storage. Additional losses are incurred when you're using an inverter to convert the DC stored in the batteries to run AC appliances. Maybe only 70 percent of the energy delivered to the batteries in a battery-charging system is eventually used productively.

To summarize: You can use the swept area of a wind turbine to quickly gauge approximately how much electricity it might produce. You can use the manufacturer's power curve to refine your estimate of the potential annual energy production—assuming you have an accurate power curve, preferably one verified by an independent testing laboratory. Or you can simply consult the manufacturer's estimate of what its turbine will do under average wind conditions.

Siting

4

One of the most challenging aspects of using wind energy is finding a place to put the wind turbine and tower—a place that's acceptable to you, your family, and the community at large. If you put a household-size wind turbine too close to the house, the turbine will suffer from the building's interference with the wind. If you put it too far away, the cost of cabling rises prohibitively, as do wire losses from electrical resistance. Every installation is a balance of these factors. Rarely is there an ideal site.

Figure 4-1. Well-Exposed Wind Turbines. Here are two examples of well-exposed wind turbines. On the left is a cluster of three Windmatics in open rangeland near Bozeman, Montana (there's not a tree for miles around). On the right, WindShare's Lagerwey turbine in urban Toronto.

But before we launch into general guidelines, it's important to note that a small book on the basics of wind energy can't begin to describe the host of issues surrounding every possible application of wind energy—from the siting of micro turbines to the development of a small community-owned wind farm.

This chapter will introduce only a few of the considerations on siting wind turbines to best advantage as well as introduce the topic of how to work with wind turbines safely. Though wind turbines by and large are safe to operate, these machines are not toys. Working with them is like working on a car 30 meters (100 ft) in the air. A mistake can kill or maim—and has!

Tower Placement

There are two primary rules for siting wind turbines. First, it should be as well exposed to the wind as possible. That is, the turbine should not be sheltered behind trees, buildings, or other obstructions. Second, as tall a tower as is practicable should be used.

Ideally, the turbine and tower should be well clear of any buildings and obstructions. If there's a hill nearby, place the turbine on the hill, even if it means a longer cable run (see figure 4-1, Well-exposed wind turbines).

It's also important that the anchors and guy

cables for guyed towers be well outside traveled ways—roads, vehicle tracks, and footpaths. If there are any farm animals roaming the site, guyed towers and their anchors must be fenced or otherwise protected.

Tower Height

Midwest wind guru Mick Sagrillo likes to warn attendees at his wind energy workshops that the top three problems owners of small wind turbines face are:

1. Too short a tower,
2. Too short a tower, and
3. Too short a tower.

Nearly everyone who has worked in the industry for a long time agrees.

Jason Edworthy has more than two decades of experience installing and operating wind turbines in Canada. His experience convinces him that the old 30-foot (10-m) rule of thumb is sacrosanct for small wind turbines. This classic axiom from the 1930s dictates that for best performance, small wind turbines should be at least 30 feet (10 m) above any obstruction within 300 feet (100 m) of the tower.

The reason: The tower needs to be tall enough to clear the zone of disturbed flow around buildings, trees, and other obstructions. This zone of turbulence can extend up to 20 times the height of the obstruction downwind and, surprisingly, up to twice the height of the obstruction upwind. Near the obstruction— your house, for example—the disturbed zone can reach twice the obstruction's height. If your house is 10 meters (30 ft) tall and you plan to install the tower nearby, the tower will need to be 20 meters (60 ft) tall. In this case, the 30-foot (10-m) rule works also. The tower should be 10 meters (30 ft) above the house, or 20 meters (60 ft) tall (see figure 4-2, Turbulence and tower height).

To avoid zoning limitations, small wind

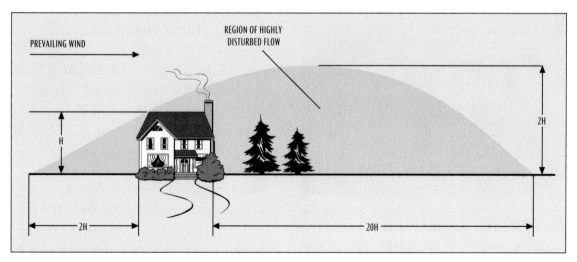

Figure 4-2. Turbulence and Tower Height. Tall towers are needed to raise wind turbines above the zone of disturbed flow around buildings, trees, and other obstructions to the wind.

Sagrillo on Tower Height

"As much as you water them, wind turbine towers don't grow," small wind expert Mick Sagrillo warns. Many wind turbine owners have woken up one morning to find that the trees they planted years before have grown, while their wind turbine has steadfastly remained at the same height.

"It's the landscape that determines tower height, not what the manufacturer offers, or the dealer sells, or zoning laws allow," says Sagrillo.

Figure 4-3. Too Short a Tower. This Bergey 1500 is on too short a tower even for windswept Patagonia. For best performance, a small turbine such as this should be installed on a tower 20–30 m (60–100 ft) tall.

turbines are sometimes installed in Europe on what North Americans would consider extremely short towers. While Europe is justifiably renowned for its development of large commercial-scale wind turbines, it has been far less successful with small turbines. European reliance on short towers may be partly to blame.

Manufacturers, dealers, and buyers in North and South America have sometimes tried to cut corners by using short and—they argue—inexpensive towers (see figure 4-3, Too short a tower). Fortunately, there has been significant progress in the development of low-cost guyed towers for micro and mini wind turbines. Several suppliers provide kits for guyed, tilt-up towers. The homeowner provides the steel pipe needed for the mast. There's now no excuse for using a tower too short for the job.

Even more promising has been adaptation of lightweight anemometer masts for use with micro and mini wind turbines. These masts have become the de facto standard among professional meteorologists and wind prospec-

tors around the world as an inexpensive and easily erectable mast system. The tubing used in these masts is not as thick as steel pipe and must be carefully matched to the wind turbine it will be supporting.

Some suppliers provide guyed, tilt-up towers

as a complete system, including base plate, gin pole, screw anchors, and guy cable. Often the guy cable comes sized with cable clamps already swaged in place. The tower sections slip together, simplifying assembly.

With the advent of inexpensive tower kits and the introduction of guyed, tilt-up mast systems for small wind turbines, there never have been greater opportunities for installing towers of the proper height for micro and mini wind turbines, those most likely to be installed by homeowners.

Mick Sagrillo agrees with Jason Edworthy about the importance of using tall towers. To emphasize this, Sagrillo says, "If it's less than 30 feet, don't bother" installing a wind turbine. He recommends towers at least 60 feet (19 m) tall for micro turbines, towers at least 80 feet (24 m) tall for turbines up to 1.5 kilowatts, and towers at least 100 feet (30 m) tall for household-size wind turbines.

There are some conditions where shorter towers may be acceptable. In electric fence charging, for example—an application in which micro turbines compete directly with photovoltaics and the turbine may be moved frequently—a shorter tower makes sense.

At sites in Scotland, Proven turbines are installed on freestanding towers only 6.5 meters (20 ft) tall. Hugh Piggott finds that in open terrain these work acceptably. Equally important, these short towers more easily get planning approval in Britain than taller towers.

To cut corners, some homeowners have mounted their micro turbines on the roof. Yes, it was done occasionally in the 1930s with the old Zenith windchargers as a marketing ploy.

But times were different then. Homesteaders were glad to simply have a charged battery on hand so they could listen to the radio or read a book by the light of an incandescent lamp. Our expectations are much greater today. We want trouble-free electricity, and rooftop mounting won't provide it.

Rooftop Mounting

Rooftop mounting of wind turbines, no matter how small, remains controversial. Southwest Windpower first suggested installing micro turbines on rooftops when it launched the Air 303, the first in their series of well-known micro turbines. Its intent was to place the Air turbines in direct competition with the simplicity and ease of mounting photovoltaic panels. Its marketing campaign was a radical in-your-face contravention of all that had gone before on the proper installation of wind turbines. Mick Sagrillo still sputters with indignation when he recalls the launch of the Air 303 as the "Air module" to emulate solar PV "modules."

In brief, all wind turbines vibrate, and they transmit this vibration to the structure on which they're mounted. All rooftops create power-robbing turbulence that shortens a wind turbine's life. Even if you were able to design a sophisticated dampening system that isolated the wind turbine from the structure, you couldn't avoid the turbulence without a tall tower rising well above the rooftop.

Installers with hands-on experience advise against it. Hugh Piggott, who lives and works with small wind turbines at his home on the windswept Scoraig peninsula of Scotland, agrees. "I wouldn't like to have one on my roof."

Mounting wind turbines—of any kind—on a building is a very bad idea. In three decades of modern wind energy, there is still no application where this has worked successfully for an extended period. Invariably, rooftop turbines are either "tied off" or otherwise inoperative (see figure 4-4, Rooftop mounting).

Promoters make two arguments for mounting wind turbines on rooftops.

- Rooftop mounting eliminates the need for a costly tower.
- Rooftop mounting enables urban dwellers to use wind energy.

Anyone who argues that they can eliminate the cost of the tower by installing a wind turbine on the roof either doesn't know much about wind energy or is at best disingenuous. This is a classic example of trying to save money by cutting corners. It's simply not worth the trouble. It will cost you more in the long run.

To avoid rooftop turbulence, the wind turbine must be raised well above the roofline by being installed on a tower. This often negates any potential savings of eliminating the tower, and increases the complexity of mounting the wind turbine and installing it safely.

No country has seen more feverish interest in rooftop-mounted or "building-integrated" wind than Great Britain. This largely results from a fundamentally flawed renewable energy policy that doesn't enable the public's use of solar energy or commercial-scale wind energy. As in Denmark during the 1970s and 1980s, the public is way ahead of their political leaders. People

Figure 4-4. Rooftop Mounting. Typical of rooftop turbines, the rotor is tied down on this nonoperating Allgaier derivative in Holzhausen, Germany in 2005. There was a similar installation of the same type of wind turbine along the main rail line near Bonn—and it, too, was tied down in the 1990s.

want to take action and develop renewable energy themselves. That the British public is frustrated, and turns to what on the surface makes sense—mounting a wind turbine on their rooftops—is understandable. Unfortunately, the public falls prey to promoters in search of the next hustle and academics in search of the next grant.

Widespread complaints of poor performance from rooftop turbines installed under Britain's ill-conceived subsidy program finally led to a series of studies and actual field trials by Loughborough University. Specialists in wind engineering, Loughborough University researchers modeled the potential performance of small rooftop turbines and concluded that they would see only one-half the average wind

rooftop promoter puts it, some installations were hard-pressed to break even between the energy produced by the wind turbine and the energy consumed in the inverter. (The turbines in the field trial were all connected to the grid.)

In the carefully guarded language of academia, Loughborough University concluded that "yields in urban areas are likely to be low." Translation: Small rooftop turbines won't produce a lot of electricity.

---

speed found at the nearest airport, for example. Keep in mind that cities are typically built in sheltered, low-wind areas. Thus, these rooftop turbines will see one-half of the typically low winds found around large urban agglomerations. And remember, cutting wind speed in half reduces the amount of power available by eight times.

In the carefully guarded language of academia, Loughborough University concluded that "yields in urban areas are likely to be low." Translation: Small rooftop turbines won't produce a lot of electricity. For example, researchers estimated that a 1.5 kW turbine installed on a rooftop in West London would produce a paltry 250 to 650 kilowatt-hours per year—hardly worth the trouble of getting planning permission, never an easy task in Britain.

And it goes downhill from there. Loughborough's field trials found that small rooftop or building-mounted wind turbines were producing not only just a fraction of the electricity estimated from wind resource models available on the Internet, but also significantly less than that estimated from the actual measured wind speed at the site.

Rather than "breaking the paradigm," as one

In some cases, there was hardly any net energy produced after accounting for that used by the inverter!

The conclusion, says David Sharman of Ampair, a manufacturer of micro and mini wind turbines, is that rooftop urban wind "can be done in some locations, but it often doesn't make economic or carbon-saving sense" to do so. To paraphrase Shakespeare in Richard III, "A tower, a tower, my kingdom for a proper tower."

Worse, a rooftop-mounted turbine can provide a nasty surprise, as an owner in upstate New York learned one stormy night when his turbine plunged through his roof. That was the end of his experimentation with rooftop mounting—and nearly the end of his marriage.

Few who consider rooftop mounting ask whether the building can support the loads created by both the wind turbine and tower. The wooden roofs of homes in North America can't support more than a micro turbine at best. A reinforced concrete roof on a commercial or industrial building might be able to withstand a slightly larger turbine.

Can the roof, then, handle the dynamic loads—the vibrations that the tower will transmit to the structure? If the building is an unoc-

Rooftop Wind in Inaction

While hiking in the southern Sierra Nevada, I was surprised to find the controversy over rooftop wind turbines dogging my steps. As we came to the crest of Bald Mountain in the heart of Sequoia National Forest, my eye was drawn immediately to the solar panels—and then to the wind turbine mounted nearby on a railing. The turbine, a Southwest Windpower Air, was not operating. As we came closer I could tell that it was tied off. *Typical,* I thought. Most of the rooftop turbines I've seen are either inoperative or simply tied off.

The fire warden came out on the platform and hollered hello as we came up the trail. I yelled back a greeting and then asked her why the wind turbine wasn't working, as there was plenty of wind on the summit. She shouted back, "So I don't have to listen to the damn thing."

Enough said. Just another example of improper wind turbine installation and the foolishness of mounting wind turbines on buildings—even a remote watchtower. —PAUL GIPE

Figure 4-sb1. Rooftop Wind in Inaction. This is an example of the improper installation of Southwest Windpower's Air module on a fire watchtower in 2006. Note that the turbine is tied off so that it won't operate.

cupied warehouse, the vibrations won't bother anyone, but if it's an office building, they may prove annoying. These dynamic loads can also damage wood frame structures that are nailed together. Mick Sagrillo likes to tell his students the story of a Bergey 850 installed on a garage in Colorado. He was asked to replace the turbine after it began pulling the roof apart. His solution was to take the turbine off the roof and install it properly on a guyed lattice mast.

The greatest hazard of rooftop mounting is caused by turbines that have not been thoroughly tested on open windy sites. We have no idea how these untested turbines will perform under load in the extremely turbulent conditions found on rooftops. No one wants to see a video on YouTube.com of a rooftop wind turbine falling off a building or destroying itself in an urban setting.

Should people living in urban areas, those in high-rises and apartment blocks, have an opportunity to invest in and develop wind energy for themselves and their community? Absolutely. Rooftop wind, however, is the wrong path to take. True urban wind can make a bold statement, produce a significant amount of electricity, and be a model for others to follow. The Danes, Germans, and Dutch have been so successful at urban wind, they don't see anything odd in it. But they learned long ago that putting a wind turbine on a rooftop was not "urban wind"—and it certainly was not "building-integrated" wind.

Building-Integrated Wind

Related to the interest in rooftop mounting of small wind turbines is the interest in integrating wind turbines into the building structure itself. This is a favorite sport of architects trying to make a name for themselves by embedding a wind turbine into their flashy architectural wonder. It may never be built— and nearly always isn't—but it still makes a big splash in the news media and allows the architects to paint themselves green—and jack up their fees.

right direction. But when it's not, what then? Turn the building to face the wind? Not likely.

Yes, we can dampen the vibrations from rooftop machines as we do with the heating and air-conditioning systems frequently mounted on rooftops. We do it all the time. But to deliver anything more than a token amount of electricity, the wind turbine has to be large relative to the size of the building. Therein lies the problem. If the turbine is large, then the loads and hence the amplitude of the vibrations it produces are large, too.

If architects want their buildings to become greener, they should cut the embedded energy (the energy used in building materials like steel and concrete), and the energy used for heating, cooling, and lighting. If that's not enough and they want to win more green points, they can always put solar panels on the rooftop. And in countries like Germany and France, which pay tariffs specifically for this purpose, architects can profitably integrate solar panels into

---

To paraphrase a Danish bumper sticker of the 1980s: Building integrated wind? No thanks. Building-integrated solar? Yes please.

Will they work? As with simpler rooftop installations, the answer is technically yes, they can be made to work. Will they work for a long time, produce the amount of electricity touted at the cost quoted? It's doubtful. Worse, they're likely to drive the occupants, then the owners, and finally the architects themselves mad with the vibrations inherent in a big spinning machine coupled to the building's structure.

Yes, there might be some speed-up effect from the building itself—when the wind is from the

the building's facade. The south side of those big glass boxes you see in fast-growing suburbs, in Germany and now in France, are covered in solar cells. When you can do that, you don't need a wind turbine.

It's noteworthy that in countries that are rapidly developing renewable energy, like Germany, no one bothers with concepts like integrating wind turbines into buildings. Sure, a drawing might appear on the occasional book cover to draw attention to the subject of renew-

Figure 4-5. Urban Wind Turbine. A reconditioned Vestas V27 at Cleveland's Great Lakes Science Center. The Cleveland Browns football stadium, not visible, is to the left. Cleveland's urban wind turbine joins a limited number of counterparts in the Anglophone world: Wellington, New Zealand's, V27 overlooking Wellington Harbor; Toronto's WindShare turbine on Lake Ontario; and the Brotherhood of Electrical Workers' Fuhrländer turbine in Boston. These are all real urban wind turbines—and they work.

able energy. But the industry and the policy makers don't take it seriously. They are too busy installing real electricity-producing wind turbines to be distracted by an architectural sideshow.

To paraphrase a Danish bumper sticker of the 1980s: Building integrated wind? No thanks. Building-integrated solar? Yes please.

Urban Wind Turbines

Intuitively, the concept immediately makes sense. Most people live in cities. Most people want to use wind and solar energy. Thus there must be ways for those living in cities to use wind energy. Indeed there is. It is being done all the time. Businesses, community groups, cooperatives, and municipal utilities throughout Europe install commercial-scale wind turbines within city limits. The idea is even catching on in North America (see figure 4-5, Urban wind turbine).

Large urban wind turbines are quite common in Germany and Denmark. However, it is much less common in Great Britain and North America. Why is that? Simple, really. Policies in much of the English-speaking world have not made it possible for urban wind turbines to earn a sufficient profit to justify the investment.

As a consequence, there are many designs in Canada, the United States, and Britain for small rooftop wind turbines that their promoters dub "urban turbines." While some are real wind turbines built by competent—but misguided—engineers, many are, for lack of a better term, simply fantasy wind turbines.

This phenomenon is being driven by the near-universal desire to participate in the renewable energy revolution sweeping the globe. Everyone wants to take part. Yet in Britain, Canada, and the United States this is often very difficult. For most residents of these countries, seemingly the only way to use wind energy is to buy small wind turbines and install them on rooftops. Promoters understand this—and prey on good citizens who want to do their part.

In continental Europe more people—including urbanites—can participate in renewable energy by investing in large, commercial-scale wind turbines installed in their community or nearby. This approach makes far more economic and environmental sense than rooftop mounting of small wind turbines.

Why? Most, if not all, rooftop wind turbines simply don't work. Yes, they may rotate, a few may even produce some electricity on occasion, but almost none works reliably or for any extended period of time. Further, most of the Internet wonders billed as "urban wind turbines" or "urban turbines" are simply a promoter's pipe dream, and few have ever been installed and even fewer—if any—are in regular service.

Fortunately, there are real urban wind turbines in North America. These are not Internet wonders—and they're not mounted on rooftops. These are real, electricity-producing wind turbines that make a difference. They can be found in Hull, Massachusetts, in Toronto, Ontario, and in Cleveland, Ohio. More are surely on the way.

For more on these turbines, see

- Hull, Massachusetts: www.hullwind. org,
- Toronto, Ontario: www.windshare. ca, and
- Cleveland, Ohio: www.glsc.org.

Noise

Wind turbines are machines operating in the wind atop tall towers. As such, they can be audible for some distance. The noise from wind turbines, though noticeable, is seldom objectionable. Yet there are noteworthy exceptions.

For their size, small wind turbines are noisier than large wind turbines, and are often placed closer to where people will hear them than large commercial machines that must meet more stringent setback requirements.

Wind turbines are noisiest when governing in high winds. Some are noisier than others. Southwest Windpower's first micro turbine, the Air 303, was notoriously noisy in high winds. Some owners likened it to the wailing of a banshee. Fortunately, Southwest Windpower's Air Breeze, the newest model in its line of micro turbines, is greatly improved and makes for a much better neighbor than the company's earlier models.

Whether you find the noise from a wind

turbine bothersome or not depends upon the model, the wind speed, how close you are to the turbine, and whether it's your machine or not.

Noise will decrease with increasing distance from the wind turbine. The simplest means of reducing noise at your home is to move the turbine as far from you as is practicable. This often is also the best solution for minimizing the turbulence affecting your turbine and for improving energy capture. This is a win–win situation.

Though you may find the noise from a wind turbine distracting at first, you could just as easily grow fond of it over time. The realization that the turbine is working hard on your behalf can gradually persuade you that its onetime whirring now sounds like a cat's soft purr. Your neighbors may not be quite as tolerant, unless they, too, share in the ownership and benefits from the wind turbine.

Birds and Bats

Despite the well-publicized problem among the thousands of wind turbines in the Altamont Pass, there's little data on the impact that single commercial-scale turbines, small clusters of commercial wind turbines, or household-size wind turbines have on birds and bats. It's reasonable to assume that wind turbines of any size or shape will kill some birds and bats in proportion to the turbine's swept area—its relative size—and the total number of turbines (see figure 4-6, Birds and small wind turbines).

There are literally hundreds of studies on the effects of wind turbines on birds and bats, many of them available on the Internet. There is also a lot of scurrilous misinformation and anti-wind propaganda. Look for reports that are written by professional ornithologists with no particular ax to grind. A number of peer-reviewed studies can be found on the Web site of the National Wind Coordinating Collaborative at www.national wind.org/publications/wildlife.htm.

Zoning and Community Relations

It's always a good idea to talk to your neighbors if you want to develop a wind project, whether it's a micro turbine or something bigger. The community and your neighbors do have a say in whether you can install a wind turbine and how you can go about doing it.

If a wind turbine is not a permitted use under the zoning laws in your area, you will need some form of zoning approval. This takes different forms from one region to another. It could be a conditional use permit or a variance from zoning regulations. And if your neighbors are not supportive, the process can be agonizing and expensive.

Zoning approval is just one hurdle. The installation must also meet all applicable building and electrical codes. If the turbine is interconnected with the grid, the utility company will have its own approval process for connecting to its lines.

While your neighbors may have legitimate concerns about wind turbines, their questions can quickly turn into fear if they are not answered promptly. Rational discussions of the benefits and impacts of wind turbines on the community can rapidly degenerate into shouting matches if the myths about wind energy and

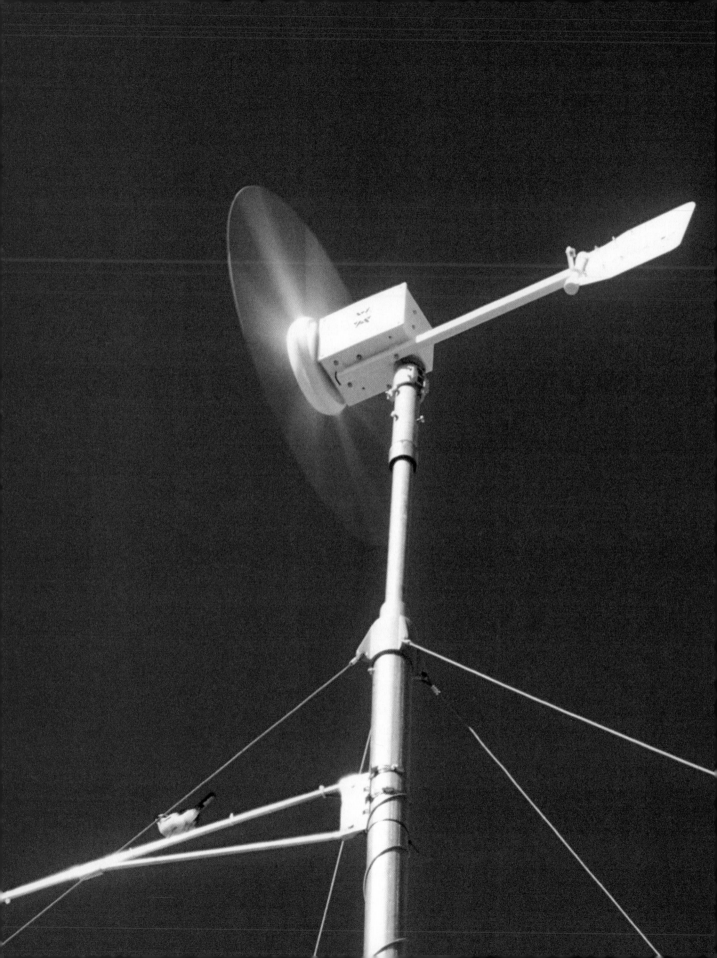

wind turbines are allowed to spread through the community without rebuttal.

Turbine Envy

Most of the wind turbines installed in North America are in commercial wind farms built by commercial developers. These wind project developers lease land from individual landowners. They lease land to install their wind turbines or ancillary structures such as roads, transformers, substations, and power lines. Developers then pay the landowners rents or royalties for the structures they place on their land.

In North America neighboring landowners without wind turbines or ancillary structures typically receive little or no payment. Though most leases in northern Europe include clauses for revenue sharing among neighboring landowners without wind turbines, this is not yet common in the Americas.

In these circumstances, one landowner receives a financial benefit, and another receives none, or very little, and yet both must live with the sight and sound of the wind turbines in their midst. The frustration at not receiving any payment whatsoever has led to "turbine envy" among some neighbors of wind projects.

Turbine envy, not surprisingly, has led neighboring landowners, who feel left out or cheated out of potential revenues, to oppose some wind projects. This phenomenon was first described in Europe, where parcels are smaller and the countryside more densely populated than in much of North America.

Down at the local pub in Britain's Cemaes Bay, for example, neighbors of EcoGen's Rhyd-y-Groes project grumbled: "Why couldn't they have put them on my land?" The wind-turbine-poor farmers say they could use the money. Under the system in existence in the early 1990s, adjoining landowners had to endure the impact without gaining any of the direct benefits.

Engineer Henning Holst faced the same problem in northern Germany. He sought to construct a royalty system based on impact zones. Those with turbines on their land would receive compensation as they do today. But turbine owners could also pay peripheral landowners more modest sums based on the parcel's distance from the project. Holst envisioned compensation based on a series of concentric zones around the turbines. Those nearest the turbines would receive more than those farther away.

Other German developers found another way: They created pooling arrangements among neighboring landowners. In one example near the central German city of Paderborn, wind developers created a land association composed of all the neighboring landowners. The land association then leased its combined land to several investor-owned cooperatives (Bürgerbeteiligungen). In this way each landowner in the pool received some form of compensation for the wind project, even if a wind turbine was not installed on his or her specific parcel.

Some wind developers in Canada have proposed similar arrangements, where neighboring landowners receive a portion of the

---

Figure 4-6. Birds and Small Wind Turbines. A bird perches on an anemometer boom beneath a Marlec 910F at the Wulf Test Field. As this scene illustrates, small wind turbines are not immune to concerns that they pose a hazard to birds.

royalties based on how much land they have in the pool. Landowners with wind turbines or other structures receive payments in addition to the payment for the lease of land in the pool.

The former minister of energy for the Canadian province of Prince Edward Island, Jamie Ballem, adapted the German pooling model for use in North America. For landowners in the vicinity of a 30 MW wind plant on the eastern end of the island, Ballem offered royalties based on distance from the turbines.

Ballem divided the royalties into three zones. Landowners with turbines on their property receive 70 percent of total project royalties. Those within 100 meters (330 ft) receive 20 percent. Those within 300 meters (1,000 ft) receive 10 percent. Total royalties from the project are 2.5 percent of the gross project revenue.

While in no way assuring that all landowners will be amenable to such an agreement or that some won't be resentful about receiving less payment than others, land pooling does increase the acceptance of renewable projects. Overall, pooling arrangements are more equitable to all landowners in an area than are leases with only some of the area's landowners.

Others have tried different approaches. German wind pioneer Heinrich Bartelt reports that he typically pays 5 percent of gross project revenues in total royalties—double the royalty payment on Prince Edward Island. In his projects in Sachsen Anhalt, Bartelt sets aside 1 percent of this 5 percent for the villages nearest his projects. Of this, a portion is set aside for community groups, sports halls, and so on. Of the 4 percent remaining, half is paid to landowners (2 percent) with turbines on their land, and half is distributed among surrounding landowners.

French project developer Erélia follows a similar approach. It pays landowners with wind turbines on their land 70 percent of the royalties paid for land leases in its Le Haut des Ailes project in northeastern France. The remaining 30 percent is paid to adjoining landowners who are not fortunate enough to have a turbine or other structure on their own land.

To spread out the royalty benefits to as many landowners as possible, Erélia moved some of the turbines slightly from preferred locations to neighboring properties, so that more landowners had turbines on their land. Similarly, Erélia worked with local landowners to avoid dividing the 30 percent of royalties into too many portions. It made preferential payments to those farmers with adjoining property or those within 80 meters (250 ft) of each turbine. It also made payments to landowners for the passage of roads and cables.

Throughout the project, Erélia kept all property owners informed of who would receive turbines, roads, and other structures. This contrasts markedly with the secrecy common among royalty transactions in Canada and the United States. Erélia also told each property owner what all other property owners would be paid—although this information was not made available to the public.

As in Germany and on Prince Edward Island, the intent of Erélia was to make the process of awarding land lease payments as open and as transparent as possible. In this way, neighbors became an integral part of the project, even though they didn't develop, build, or own it themselves.

Safety

Of the 36 people killed working around wind turbines worldwide in the past 30 years, 3 have been killed working on small wind turbines. There have been numerous close calls. Here are a few suggestions for working safely with the wind. For more details see *Wind Power: Renewable Energy for Home, Farm, and Business.*

Moving Machinery

Wind turbines contain rotating machinery, and every warning about how to work around such machines applies. Don't wear long hair, loose clothing, rings, or necklaces around any turning shaft—regardless of how slowly it's turning. Hugh Piggott in *Windpower Workshop* recounts a "hair raising" tale told by Mick Sagrillo of the encounter between Mick's ponytail and the slowly turning shaft of a "Jake" in an Alaskan shop. (If you have long hair, at least keep it tucked into your shirt or under your hat.)

Never go near a spinning wind turbine.

Period. If you have to work near the turbine, furl it, pitch the blades, or otherwise turn the turbine off and brake the rotor to a stop. (If the wind is light to moderate, you can bring most rotors driving small wind turbines with permanent-magnet generators to a halt by shorting all phases in the armature.) Then place a locking pin into the drive train so it can't turn, and lock the turbine so it can't yaw about the top of the tower. Unfortunately, few small wind turbines include such fundamental safety features.

Electrical

Permanent-magnet alternators produce current whenever the rotor turns—even when disconnected from the load or control panel! Before servicing the control panel, disconnect the power supply from the turbine. Even though you may be using a low-voltage DC power system, many wind turbines produce three-phase AC and when unloaded can reach very high voltages. Install a fused disconnect switch for this purpose. It will come in handy.

Tower Work and Do-It-Yourselfers

Any wind turbine and tower that cannot be safely lowered to the ground for servicing should have a fall-arresting system for ascending, descending, and working atop the tower, a sturdy work platform, and safe, clearly identifiable anchorage points for attaching your lanyard. Never climb a tower of any type unless you've received training in tower safety.

Homeowners should only attempt installing wind turbines less than 3 meters (10 ft) in diameter on lightweight tilt-up guyed masts. Homeowners should avoid installing larger turbines, freestanding truss towers, or heavy-duty guyed towers without hands-on training. Most lack the skills, specialized tools, and safety equipment necessary. The tools can be purchased, and the skills needed can be learned. Workshops, such as those that Mick Sagrillo teaches, or installer training programs offered by manufacturers, are worth the money and are the best way to learn how to install wind turbines safely. A book is no substitute for the hands-on learning that's required.

Necessary Steps for a Small Wind Project

The following has been adapted from a summary by Jim Green at the National Renewable Energy Laboratory in Boulder, Colorado, on the steps necessary for the successful installation of a small wind turbine. The list applies to commercial wind turbines as well, with the additional step of studying their environmental impact.

- Assess your electricity consumption, the cost, and your utility tariff.
- Reduce your consumption through conservation and improved efficiency.
- Estimate or measure your wind resource.
- Select turbine size, model, and tower height needed.
- Determine incentives—if any—and estimate the economics.
- Notify your neighbors and apply for zoning where required.
- Obtain an interconnection agreement from your utility.
- Obtain a building permit.
- Order the turbine and tower.
- Install the turbine safely.
- Commission the turbine and make sure it's working properly.
- Periodically inspect and maintain the turbine and tower.

Off-the-grid power systems can experience high current draws and high charging rates. Both conditions require that all cabling be amply sized and the connections terminated correctly, for safe operation.

Fuse all power sources (both wind and solar, for example), both AC and DC loads, and connections to the batteries. There are many pre-engineered and assembled panels that include built-in fuses or circuit breakers for both the DC and AC side of off-the-grid power systems. These are available under various trade names. Some are approved by various standards or rating organizations, such as Underwriters' Laboratories, and will pass muster with building inspectors in much of the United States. They are part of an encouraging trend toward more standardized and professional DC to AC power systems. Use them.

If you have any doubts about how to properly fuse a part of your power system, or how to make sound terminations, consult the manufacturer or supplier of the component, or hire a licensed electrician.

Batteries

Always use extreme caution when working around batteries. Use the same precautions you would use when working near an automotive battery. Do wear goggles to protect your eyesight from spraying battery acid should an accident occur.

Beware of dropping metal tools onto exposed battery terminals. This is a recipe for disaster. Some pros recommend insulating metal tools for working around batteries.

Vent the batteries adequately to the outdoors to prevent concentrations of explosive hydrogen gas. Avoid any source of sparks or open flame around the batteries.

Maintenance of a medium-size wind turbine. Technician servicing a Vestas V27, a 225 kW wind turbine, atop Alta Mesa near Palm Springs, California. Working with modern wind turbines requires professional training—and a desire to "see the world."

> ### *"The life of a wind turbine is directly related to the owner's involvement." —Jim Sencenbaugh*

As when working around rotating machinery, don't wear rings or necklaces around batteries.

Towers

Working on or around wind turbines and their towers poses two kinds of hazards. One is a fall from the tower; the other is being hit by something falling from it.

As Hugh Piggott often notes, "bits and pieces" fall off the 30-odd turbines he maintains near his home in northwestern Scotland. So never work beneath an operating wind turbine. Turn it off or brake it to a stop before performing any service at the base of the tower. Even a small nut can pick up a lot of speed from 20 meters (60 ft) overhead and cause a nasty injury if anyone is unlucky enough to be in its way.

Ideally, guyed towers should be located two tower-lengths away from occupied buildings and power lines. The reason for this may not be apparent. It's clear that you or someone else could be injured by the mast if it fell over. What's not so obvious is that the guy cables can go slicing through the air well beyond the end of the mast. The whipping guy cable could not only cause serious injury but also lash a bare conductor on a nearby utility line, making the whole tower and all its guy cables "hot."

Servicing

All wind turbines must be regularly inspected from time to time and serviced, as well. Mick Sagrillo likes to quote Jim Sencenbaugh, a highly regarded wind turbine designer of the 1970s, on the need for routine inspection and service: "The life of a wind turbine is directly related to the owner's involvement."

It's important to anticipate how you will get to the wind turbine. Working atop a tower is always dangerous. "Avoid it if at all possible," says Scoraig Wind Electric's Piggott.

For micro and mini wind turbines, use a tilt-up tower whenever possible.

But if you have to use a freestanding tower, climbing may be your only option. Homeowners seldom have access to the bucket trucks or person lifts that can make servicing a small wind turbine both simpler and safer.

Mick Sagrillo recommends that dealers and homeowners who plan to service wind turbines atop a tower should take a tower safety training course before they venture forth on their first run up the tower.

Always wear and use an approved safety belt and lanyard when climbing and working atop a tower. *Home Power* magazine occasionally carries articles on how to use basic safety equipment. Belts and lanyards are also discussed in *Wind Power: Renewable Energy for Home, Farm, and Business.*

Climbable towers should always include a fall-arresting or safety cable. When ascending or descending the tower, attach your safety belt to the cable with the special sliding shoe that's provided.

A work platform should always be included on climbable towers. This need not be elabo-

Figure 4-7. Tower Safety. Remember, the turbine has to be serviced. If it cannot be lowered to the ground on a hinged tower, it must have a work platform and anchors for a fall-protection lanyard. The tower must also have an approved ladder and a fall-protection cable or rail. Note the work platform on either side of the nacelle of this Vestas V15. The access ladder is on the inside of the tubular tower.

rate, but should be sufficient to allow safe and comfortable servicing of the wind turbine. Be advised that most non-tilt-down towers made for small wind turbines in the United States don't include work platforms. Why remains a mystery (see figure 4-7, Tower safety).

Tilt-up, guyed, tubular masts must be lowered to allow servicing of the wind turbine. This entails its own set of hazards, but it does eliminate the need, at least for small wind turbines, for working on the turbine in the air.

Raising and lowering tilt-up towers is also risky because of the heavy loads on the guy cables when the tower is near the ground. Never stand underneath the tower or guy cables when it is being raised or lowered. Something unexpected can go wrong. In the dairy state of Wisconsin, windsmiths call guy cables on tilt-up towers "cheese slicers" for a reason.

An experienced field crew at the Alternative Energy Institute at West Texas A&M once was lowering a 10-meter-diameter, 25 kW turbine at their test field. They had done it many times before. But this time was different. There was miscommunication. To make matters worse, there was a photographer in the path of the tower. The tower whizzed by the photographer's head—missing by inches—and crashed to the ground. No one was hurt, but there were a lot of deep breaths and red faces. The photographer quickly left Texas, never to return.

Don't let this happen to you.

Off the Grid

5

Thirty years ago we envisioned the development of wind energy as thousands of small wind turbines dotted across the continent, supplying local needs and feeding excess power into the grid. Every farm, ranch, and rural home would have its own turbine. Sure, there would be the odd battery-charging turbine for a Third World village, or a mountaintop repeater station. But the thousands of machines that manufacturers would be turning out would be destined for delivering utility-grade electricity interconnected or "intertied" with the grid.

It hasn't worked out that way. Yes, there are parts of Denmark and northern Germany where nearly every farm does have its own turbine. But these are large commercial-sized wind turbines producing millions of kilowatt-hours per year. They sell all their generation to the utility just as they sell milk to the local dairy cooperative.

Instead, the boom in micro and mini wind turbines, especially in North America, has been in battery-charging systems. The reasons are part political, part technological. In North America, small wind turbines interconnected with the local utility simply don't pay. But equally important has been the development of inexpensive micro and mini wind turbines that make wind energy more affordable and easy to use in off-the-grid systems. Meanwhile, advances in electronic inverters, compact-fluorescent lamps, other low-power electronic devices, and photo-voltaic panels have enabled those off the grid to enjoy the same standard of living as those on it. They will have to dramatically cut their electricity consumption compared with the typical North American, but that's never been easier. Many North Americans are surprised at how easy it is to cut their electricity consumption in half (see figure 5-1, Stand-alone wind machines).

"Throughout the mountainous West," says Bill Dorsett, "there are isolated valleys where the cost advantages of powering remote cabins with solar cells are so compelling that no one thinks twice." In Kansas, where Dorsett lives, "the only geographical barrier to grid electricity is space. Yet even here we are looking to a resurgence of decentralized power," he adds. "Today there are livestock pumps, fence chargers, and certainly whole farmsteads which are remote enough to be cost-effective for stand-alone solar electric or wind power."

The photovoltaic industry has aggressively sought out and now serves the needs of a growing number of off-the-grid applications. The spin-off has been increasing interest in small wind machines, notably micro and mini turbines, for adding to existing solar systems. This improves off-the-grid performance in winter when solar's contribution is at a minimum.

Long before the development of solar cells, wind energy had been used in a multitude of

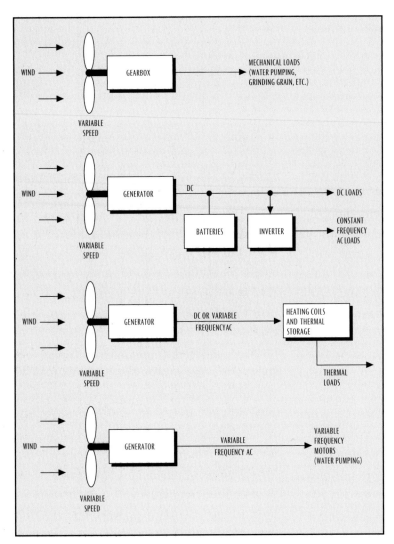

Figure 5-1. Stand-Alone Wind Machines. Micro, mini, and some household-size wind turbines can be used off the grid in a number of applications. Commonly they are used to charge batteries for providing both AC and DC loads (upper left). With modern electronic controllers, they can also drive electric well pumps and other motors directly without batteries (lower left). Small wind turbines have also been used to heat homes (upper right). In much of the world, remote homesteads and ranches still use farm windmills to mechanically pump water (lower right).

off-the-grid applications. Historically, wind turbines had been used to grind grain and pump water. In the 1930s thousands of small wind turbines were used on the Great Plains to meet the electrical needs of remote homesteads.

Recreational Vehicles

Today's small wind turbines are finding an increasing number of novel applications. Thousands of micro turbines are used on sailboats around the world, and now they're appearing strapped to the sides of the lumbering land yachts that cruise America's byways. Since recreational vehicles already have batteries and a charging circuit in place, micro turbines are an ideal addition to the 12-volt system. Many RVs sport solar panels to reduce the running time on their noisy generators, and micro turbines are a natural extension (see figure 5-2, Micro wind turbines for sailboats).

Figure 5-2. Micro Wind Turbines for Sailing. The venerable Ampair 100 can be found charging batteries on sailboats around the world, as here in Copenhagen's inner harbor.

Figure 5-3. Small Cabin Charging. An LVM micro turbine on a Danish summer cabin. Note that these summer cabins are just that: small cabins for summer use only. No one lives in this building. This is one of the few situations in which mounting a turbine on a building may be tolerable.

Cabins and Cottages

It's a simple progression from using a micro wind turbine on a sailboat or an RV to installing a micro turbine and batteries in a small cabin. These 12-volt systems often employ the same hardware and appliances found in RVs. Cottages with heavier loads than those found on RVs may opt for a more powerful 24-volt system (see figure 5-3, Small cabin charging).

Electric Fence Charging

Once the sole domain of photovoltaics, electric fence charging is a potentially significant new market for micro turbines. Electric fences are much more widespread elsewhere in the world than in North America. Denmark, for example, has been using electric fences to enclose fields since before World War II. Electric fencing is also quite common Down Under for managing sheep in New Zealand. A 12-volt micro turbine and a weatherproof battery-pack are all that's needed (see figure 5-4, Electric fence charging).

Telecommunications

One of the early applications for small wind turbines, as well as photovoltaics, was powering remote telecommunications sites. These

Figure 5-4. Electric Fence Charging. This 12-volt Marlec 910F on a guyed pipe tower is charging an electric fence on the banks of Denmark's Skibsted Fjord. Solar-powered electric fences are common in Europe, but micro wind turbines are gaining ground.

Figure 5-5. Telecommunications. A Bergey Excel is used to power a QuebecTel station at Pointe-au-Pére on the southern shore of the St. Lawrence River.

were often located on nearly inaccessible mountaintops where trucking in diesel fuel was difficult and expensive. The proliferation of new telecommunications companies and the explosion in the use of cellular phones could require installation of more wind-diesel hybrids. Like off-the-grid systems for remote homes, these hybrids also include solar panels and batteries (see figure 5-5, Telecommunications).

Wind Heating

Like wind-and-solar hybrids, using winter winds to heat your home has always seemed an ideal way to marry a technology with a natural cycle. Since heating loads are a function of heat-robbing winds, why not use those very winds to heat your house? The University of Massachusetts proposed such a Wind Furnace in the mid-1970s, and several companies have tried to market the idea to homeowners already on the grid. The concept never caught

Figure 5-6. Home Heating. The Folkecenter for Renewable Energy designed this 7 kW turbine to produce low-grade electricity for heating hot water. Despite many attempts, the market for successfully using wind turbines to produce heat has proven elusive.

on in North America, where the economics never made sense, but it did briefly flourish in Denmark, where heating prices are considerably higher.

Denmark's Folkecenter for Renewable Energy has found that a wind turbine that covers winter heating demand can easily cover domestic hot-water loads in summer. The Folkecenter also reports that it is economically more advantageous to use wind-generated electricity as electricity, rather than converting it to heat. Typically, electricity has double the value per kilowatt-hour of electricity converted to heat with resistance heaters (see figure 5-6, Home heating).

Proponents have argued that storing excess wind energy as heat is much cheaper than stor-

ing it in batteries. Thus heating with wind can play an important role in stand-alone systems that are usually remote and don't have ready access to fossil fuel. Scoraig Wind Electric's Hugh Piggott dumps his excess wind power into "storage heaters" that store heat in the form of hot water. These are common in Great Britain for storing cheap off-peak electricity at night for those on the grid, says Piggott.

To guarantee a reliable supply of wind-generated electricity for an off-the-grid system, you need a much bigger turbine than would be necessary to produce the appropriate number of kilowatt-hours for a grid-connected home. The bigger turbine eases the load on the batteries by generating current even in light winds. In moderate or strong winds the turbine produces more power than needed. The surplus can then be used for heating, thus saving fuel.

Dumping or diverting excess wind power to heating has become a common practice in stand-alone wind systems. For those off the grid, the situation in Alberta, Canada, is little different from Scotland's. In Alberta, Jason Edworthy used a Bergey Excel in a battery-charging system where excess energy was diverted to a conventional hydronic system for in-floor heating.

The turbine's electrical control system must be adapted to direct excess power to your dump or diversion load to ensure that this load is compatible with the wind turbine. You don't want to stall the rotor, when the dump load switches on, by placing too great a load on the rotor. Modern solid-state electronics are well suited for this function.

Hybrid Wind-and-Solar Systems

You might say that joining wind and solar is a marriage made in heaven. The two resources and technologies are complementary. Together they not only improve the reliability of a stand-alone power system, but also are more cost-effective than either one alone (see figure 5-7, Hybrid wind and solar).

Many off-the-grid homes start with a few photovoltaic panels, because they are simple to install and their unit costs are within reach of most homeowners. A small system may use two or more panels in series or parallel, a few batteries, and some 12-volt or 24-volt DC appliances.

Until a decade ago, adding a wind turbine to hybrid system was problematic because even the smallest turbines cost more than most people were willing to spend. The advent of inexpensive micro turbines, however, has greatly facilitated building hybrid wind-and-solar systems.

Today it's not uncommon in rural areas without utility power like Tehachapi, California's, Mountain Meadows community to see suburban homes with a mini wind turbine, a full solar-PV array, a battery shed, and a backup gasoline or diesel generator—for when all else fails (see figure 5-8, Hybrid stand-alone power system).

Because wind energy has greater power density than solar, at even low-wind sites, the addition of small amounts of wind capacity, like that of a micro turbine, can significantly

Figure 5-7. Hybrid Wind and Solar. This Proven wind turbine works in conjunction with photovoltaic panels to power a remote telecommunications site in Scotland. (Proven Engineering, www.provenenergy.com)

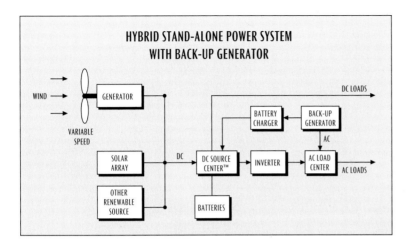

Figure 5-8. Hybrid Stand-Alone Power System. A modern battery-charging power system will use a combination of wind and solar generation, batteries, inverters, and backup power sources.

boost the total energy available. See table 5-1, Comparison of Generation by Solar and Wind; and figure 5-9, Solar yield.

"Hybrids are simpler than most people expect," says Jason Edworthy, "at least until you add a generator with automatic operation." The automation that most of us desire quickly complicates life, he warns.

When Ed Wulf was building his home in Southern California's Tehachapi Mountains, the local utility offered to bring him power for a mere $50,000. He told them, "I can do better than that." He only had to look out his picture window for the solution. His window opened onto one of the world's largest wind power plants. The thousands of wind turbines across from Wulf's homestead churn out enough electricity to serve the needs of more than 500,000 Californians. For regulatory and logis-

tical reasons, the area's wind power plants were unable to help Wulf. He had to go it alone. But the very existence of those turbines proved that wind energy would work for him.

Wulf set out to install his own stand-alone power system, a hybrid that would use the area's wind and solar energy. He installed a mini wind turbine, a large photovoltaics array, batteries, a diesel generator, and an inverter to power his conventional home. His was not a minimalist approach. Wulf's system could probably power an entire Third World village.

Components, such as batteries and inverters, are critical to the success of hybrid systems. As hybrid systems have become more sophisticated, so, too, have various components used in their design. However, batteries have always been an expensive and troublesome part of off-the-grid systems. Battery storage costs about

Table 5-1: Comparison of Generation by Solar and Wind							
	Wind Speed		Intercept Area				Max	Min
	m/s	mph	m²	ft²	~kWh/yr	~kWh/day	kWh/mo	kWh/mo
Solar PV			1	11	100	0.3	12	2.2
Wind	4.0	9.0	1	11	200	0.5	22	12
Source: Nordvestjysk Folkecenter for Vedvarende Energi, Denmark.							

Figure 5-9. Solar Yield. Tracking solar arrays, such as these outside Madison, Wisconsin, improve performance over modules that don't follow the sun. Even at low-wind sites, a wind turbine will produce significantly more electricity for an equivalent intercept area—the area of the four solar modules on these trackers—than solar PV. However, solar PV is nearly trouble-free, and small wind turbines are not. Wind and solar work better together than either does alone.

$100 per kilowatt-hour, not counting freight. And shipping heavy batteries is costly (see figure 5-10, Battery storage).

Batteries also have a limited lifetime. The Folkecenter for Renewable Energy estimates that batteries are good for at least 2,000 cycles at 50 percent discharge because you can't draw much more than 50 percent of the stored energy out of lead-acid batteries without sulfating the plates and reducing their effectiveness. A typical golf-cart battery will store 1 kilowatt-hour and will discharge about 1,000 kilowatt-hours (2,000 cycles of 1 kWh at 50 percent capacity) of energy over its lifetime. (The battery will still be usable after 2,000 cycles, but at a reduced capacity.) Thus battery storage costs more than $200 per kilowatt-hour of usable capacity.

At more than $200 per kilowatt-hour of storage and 1,000 kilowatt-hours of total storage, electricity drawn from a battery costs more than 20 cents per kilowatt-hour for the batteries alone. Because of such high costs, designers

Because wind energy has greater power density than solar, at even low-wind sites, the addition of small amounts of wind capacity, like that of a micro turbine, can significantly boost the total energy available.

Figure 5-10. Battery Storage. Mike LeBeau's tidy and well-organized battery room near Duluth, Minnesota. Inverter (left), power center with disconnect switch (center), charge equalizer (right), and batteries. Note that all cables run in conduit or raceways. Ideally, the batteries should be isolated from the electronics in a well-ventilated room.

make every effort to limit the amount of battery storage needed in an off-the-grid system by using the sun and wind together.

For RV, cabin, or all-around battery-charging systems, don't overlook the 220-amp-hour golf-cart battery. These batteries are widely available throughout the Americas. The 350-amp-hour L16 batteries from Trojan are also popular, but half again as expensive as golf-cart models.

For larger systems with many AC loads, inverters are a must, and today's electronic inverters have made living off the grid with AC appliances easier than ever before. Real Goods' Doug Pratt says that new inverters are so programmable, "They'll do whatever you want them to do." For example, they can be programmed to start certain heavy loads when excess power is available, and cut the load when battery voltage falls, or to start and stop a backup generator as needed.

Wind Pumping

One of the oldest uses of wind energy is to pump water, and that use is still important today. There are more than one million water-pumping windmills in use worldwide. But today's wind technology offers far more choices for how to pump water than were available just a few years ago: traditional mechanical wind pumps (farm windmills), air pumps, and wind-electric pumping. There are advantages and disadvantages to each.

Classic farm windmills are "hard to beat in light-wind regimes," says Eric Eggleston, formerly a researcher at the US Department of Agriculture's experiment station in Bushland, Texas. Their high-torque rotors were designed for pumping in the light winds of summer on America's Great Plains. And they do their job well (see figure 5-11, Mechanical water pumping). But whether a farm windmill is an optimal choice depends upon the wind available, the depth to water, and the amount of water needed.

Eggleston explains that although an American farm windmill costs about 10 percent less than an equivalent size wind-electric system, the farm windmill will pump only half the volume. This is due to the better performance of the more aerodynamic rotor on the modern wind-electric turbine, and to a better match between the rotor's performance and the power available in varying winds.

The siting of mechanical wind pumps is also limited. Farm windmills must be placed directly over the well, whereas wind-electric pumping systems permit the wind turbine to be placed to best advantage. An electrical cable is then used to connect the wind-electric turbine

Figure 5-11. Mechanical Water Pumping. The American farm windmill is well suited to mechanically pumping water in the light winds of summer, when water is most needed.

with a pump motor at the well. For low-volume applications, air-lift pumps also offer similar flexibility, since they use pliable plastic tubing.

Unlike earlier designs with batteries and inverters, contemporary wind-electric pumping systems drive well motors directly. The key has been development of electronic controls that match the pump motor load to the power available at different wind speeds. The control system is the limiting factor, says Nolan Clark, director of the USDA's Bushland station.

While a farm windmill side by side with the same size modern wind turbine will pump less volume than the modern turbine, it will do so when the water is needed most—in light summer winds. The high solidity of the rotor on a farm windmill ensures that it will pump water in the faintest of summer breezes. Where water for domestic or livestock needs is critical in arid parts of the world, the farm windmill will continue to have an important role.

In the next chapter we'll unravel the mystery surrounding a utility intertie. For a more extensive discussion of how to use wind energy for off-the-grid applications, see Chelsea Green's companion book *Wind Power: Renewable Energy for Home, Farm, and Business.*

Interconnection

6

The once promising prospect of tens of thousands of small wind turbines whirring above farms, ranches, and homes across the breadth of the United States, feeding electricity into the utility network, has never fully materialized. Yes, technically and legally, it can be done. And there are thousands of small wind turbines distributed across North America doing just that, but neither in the numbers nor in the manner first envisioned.

Following the oil crises of the 1970s, many consumers worried that utility prices would continue rising for decades to come. This fear provoked keen interest in using small wind turbines to offset consumption of electricity from the local utility. These wind turbines would be interconnected or intertied with the utility on the customer's side of the kilowatt-hour meter.

Electric utilities, of course, had no interest in furthering such ventures and placed numerous roadblocks in the path of anyone who tried. But when the passage of the Public Utility Regulatory Policies Act (PURPA) in 1978 removed at least the legal obstacles to interconnection, and Congress granted generous subsidies, small wind technology seemed poised for explosive growth.

In part, the effort succeeded. There were more than 4,000 small wind turbines installed in distributed applications in North America for this purpose. Similarly, California's successful wind farms are also a direct result of PURPA and Congress's tax incentives. But a host of problems beset wind turbines large and small. Some problems were technical: The turbines didn't work as well as expected. Some problems were commercial: Firms entered and left the business so fast that it was hard to tell who was a manufacturer and who wasn't. And then the price of oil collapsed, and with it concerns about rising utility prices in North America.

Shortly thereafter the North American wind industry collapsed. It was old, pedestrian uses, such as battery charging and water pumping, that saved small turbine manufacturers from extinction. Because these applications were more prosaic than feeding electricity into the utility system on a par with that from a nuclear plant, they were viewed as not quite so alluring in the American context and were very nearly overlooked. Fortunately, Europeans, especially the Danes and the Germans, kept the dream of distributed generation as part of the electricity grid alive in the "old world." Their success has now become a model for North Americans.

Times have again changed. The price of oil and natural gas has begun a steady march ever higher as scarcity is beginning to stare us in the face. Increasing numbers of homeowners and businesses are looking to small wind turbines to supplement their need for electricity through net metering. But a potentially far more significant development is the growing movement toward locally owned distributed generation using medium-size and large commercial-scale wind turbines, following the European example. (There will be more on "community wind" in the next chapter.)

Without higher electricity prices, small wind turbines seldom make economic sense in much of North America where utility power is already available. Electricity prices, despite recent hikes, remain low throughout North America relative to what others pay around the world. Small wind turbines remain costly for the amount of electricity they produce, and their reliability over the long period of time needed to justify their expense is problematic. The situation is fluid as *Wind Energy Basics* went to press in early 2009. Prices for electricity are expected to rise rapidly as the rocketing price of fossil fuels begins to be felt throughout the economy. Further, new small wind turbines are entering the market that could change the economic calculus.

Of course, it's unfair to demand that small wind turbines meet artificial economic criteria today when the future cost of electricity will certainly be higher, but by how much is anyone's guess. Few consumers think twice about whether a houseboat, a vacation cabin, or a pair of snowmobiles parked on the trailer in the backyard is economic or not. Small wind turbines, regardless of their cost, provide a necessary service, unlike many frivolous expenditures.

The American political context could change overnight. Another war in the Middle East, calls to combat global climate change, an outright shortage of oil, a Western reactor exploding in a

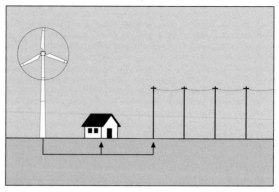

Figure 6-1. Net Metering or Self-Generation. The wind turbine is connected to the electricity distribution system on the customer's side of the kilowatt-hour meter. The wind turbine serves on-site consumption; any excess is fed back into the grid. This is seen as "negative load" by the electric utility, and "energy conservation" by policy makers.

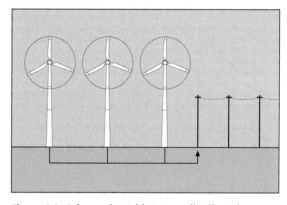

Figure 6-2. Sales to the Grid. In true distributed generation, wind turbines produce all their electricity for sale to the grid. Nearly all the wind turbines dotting the landscape of Germany and Denmark sell their electricity directly to the local electric utility through feed-in tariffs.

populous region—any of these could push the North American public and political leaders to take action.

Nevertheless, medium-size and large wind turbines—under the right conditions—can make economic sense today. That's the reason wind farms, or more correctly wind power plants, are springing up nearly everywhere. But wind energy is more than just small wind turbines offsetting domestic consumption (see figure 6-1, Net metering or self-generation). It is more than giant wind power plants producing bulk electricity for transmission to distant cities. Wind energy today is also the use of large wind turbines singly or in small clusters connected to the grid, delivering their electricity to the local distribution system—for a profit (see figure 6-2, Sales to the grid).

Interconnection Technology

The distributed use of wind energy entails connecting the wind turbine either on the customer's side of the kilowatt-hour meter, or directly to the utility's distribution system through a dedicated meter and transformer (see figure 6-3, Interconnected commercial wind turbine).

In the early 1980s most small wind turbines designed for utility interconnection used induction generators just like the commercial machines then being installed in California's windy passes. A few, like the Bergey Excel, were a sign of things to come and used permanent-magnet alternators feeding electronic inverters that provided the utility-grade electricity needed in the household. Today most small turbines destined for the customer's side of the kilowatt-hour meter follow that model.

Induction or Asynchronous Generators
Wind turbines using induction generators are certainly the simplest to connect to the grid. Early promotions by defunct 1970s manufacturer Enertech showed a wind turbine with an electrical cord for plugging into a wall outlet. Most of the medium-size wind turbines operating in wind power plants around the world are nothing more than glorified induction generators with an electrical cord. No mysterious

Figure 6-3. Interconnected Commercial Wind Turbine. Small commercial-scale and large wind turbines may be connected on an industrial or commercial customer's side of the meter or—as with many large wind turbines connected to the utility's distribution lines—via a transformer. In the photo, the transformer is the large green box in the foreground. Note its architecture—functional yet aesthetically pleasing. This is a common practice in northern Europe.

black box is needed to convert the generator's output to the form used by the utility.

The induction generator, or asynchronous generator as it's known technically, uses current in the utility's lines to magnetize its field. Thus, the voltage and current that the induction generator produces are always synchronized with those of the utility. In principle, the induction generator is unable to operate without the utility. When the utility's system is down, the induction generator is, too.

Electronic wizards can fool induction generators into thinking the utility's line is present by using capacitors to charge the field, enabling induction generators to be used in stand-alone power systems. Vergnet wind turbines, for example, drive induction generators for use in wind-diesel and battery-charging systems in France and its overseas territories. For nearly all other applications, however, wind turbines using induction generators automatically disconnect from the grid when the utility's lines are down.

Induction wind turbines do use sophisticated electronic controls to tell the wind turbine when to connect to and disconnect from the grid. If they didn't, the wind turbine would "motor" in low winds, acting like a giant fan and consuming electricity instead of generating it. These controls also tell the generator to engage the utility's line gradually in small steps rather than suddenly all at once. Such soft starts reduce the voltage flicker that once characterized the addition of wind turbines driving induction generators on rural distribution lines.

Electronic Inverters

Years ago, it seemed miraculous when Windworks announced that you could use a synchronous inverter to connect a 1930s-era Jacobs wind turbine with the lines of your local electric utility. Although the technology for doing so had been around for some time, it wasn't widely known among alternative energy enthusiasts. The technology is much less mysterious today. The electronics are now quite commonplace, if not passé.

This inverter and the others that quickly followed took DC, in the case of the old Jacobs generator, or rectified the variable-voltage, variable-frequency AC from an alternator to DC, inverted it to AC, and synchronized it with the AC from the electric utility. In this way, old synchronous inverters were line-synchronized or line-commutated. They used SCR (silicon controlled rectifier) switches with analog controls to signal when they would feed bits of current into the utility's lines. Because they were line-commutated, they needed the utility's line present to function.

Contemporary inverters are self-commutated. They can produce utility-compatible electricity using their own internal circuitry with IGBTs (integrated-gate, bipolar transistors) and digital controls. The new self-commutated inverters greatly improve reliability and power quality over the older line-commutated versions. Advances in inexpensive inverters for solar photovoltaic systems have produced product spin-offs for small wind turbines.

Successful German inverter manufacturer SMA now markets its Windy Boy inverters for use with small wind turbines. Some suppliers of small wind turbines have designed their permanent-magnet alternators to work specifically with the SMA's Windy Boy inverters.

Other manufacturers, such as Southwest Windpower, have developed their own, purpose-built inverters. The inverter for Southwest Windpower's 3.7 model is built into the nacelle. Like the Enertech of old, Southwest Windpower wants its 3.7 model to be "plug and play." The output from the 3.7 is wired directly into the consumer's electrical service panel (see figure 6-4, Plug-and-play?).

Advances in electronics have also made large, direct-drive wind turbines possible. Enercon, the largest supplier to the German wind turbine market, uses electronic inverters in much the same manner as small wind turbines. Similarly, most other large wind turbine manufacturers, those relying on a conventional drive train with a gearbox, use doubly-fed induction generators coupled to electronic inverters.

With the exception of a few small wind turbine designs, a few small commercial-scale turbines, and a few older models of large wind turbines, the wind industry has steadily moved to generator–inverter combinations for producing utility-compatible electricity.

Power Quality and the Utility

Always consult your local utility before attempting to interconnect your wind turbine with its

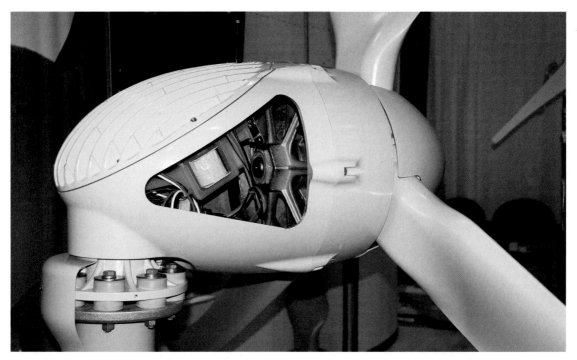

Figure 6-4. Plug-and-Play? Southwest Windpower's Skystream incorporates an inverter built into its nacelle. Note the heat sink and cooling fins on the left of the nacelle. The firm's objective is to make the electrical installation as simple as possible. The downside is that when there's a problem with the inverter, the wind turbine has to be lowered to the ground.

lines. The utility has valid concerns about the safety of your installation and the quality of the power that will be produced. It will be interested in the power factor, voltage flicker, and harmonics produced by your wind machine. It will also have concerns about the safety of its personnel when working in the neighborhood. These concerns are no reason to deny

This explicitly states that renewable sources of generation, if they meet reasonable safety and power-quality requirements, have a priority for connection to the grid, and the sale of the electricity produced has priority over the utility's other sources. Unless the utility can show clear technical reasons why it should not connect, it must allow the interconnection.

Wind turbines have operated more than 10 billion hours on the lines of electric utilities in Europe, the Americas, and Asia without bringing the world to an end.

---

an interconnection, but the utility has a right to ask these questions and to get the answers before accepting an interconnection with their lines. The utility may also require payment for reasonable costs arising from the interconnection. This shouldn't be much, but get it spelled out beforehand.

Unfortunately, even after more than 30 years of modern wind energy, it may take an attorney and a big bank account to convince the utility to do what it's obligated to do. Wind turbines have operated more than 10 billion hours on the lines of electric utilities in Europe, the Americas, and Asia without bringing the world to an end. They have generated some 1,000 terawatt-hours (1,000 billion kilowatt-hours) of clean, nonpolluting electricity worldwide. That should be sufficient proof that wind energy can work harmoniously with the existing electricity distribution system.

Germany's impressive development of wind energy, much of which is distributed, is due in large part to its groundbreaking Renewable Energy Sources Act, the subtitle of which is "The Act on Granting Priority to Renewable Energy."

Sadly, this was the original intent of PURPA in the United States. PURPA grants the right to interconnect, but the act has often been thwarted by recalcitrant utilities trying to protect their monopoly franchises. The situation in Canada and Mexico, where there is no equivalent to PURPA, is even worse. As a consequence, wind energy advocates have been forced for decades to plead with legislators and cooperating utilities for the ability to connect. The result has typically been inadequate—in the form of net metering, where the wind turbine can only be used on the customer's side of the meter to offset domestic consumption. Even today there are states and provinces that don't allow net metering.

Degree of Self-Use

In states and provinces where consumers are not allowed to run their kilowatt-hour meter backward, it becomes critically important to know how much of the wind generation will be used to offset on-site consumption.[5] This is the degree of "self-use," or generation used on-site (see figure 6-5, Degree of self-use). Output from a wind turbine varies with the wind. Electrical consumption in your home or business varies with the time of day. When you try to match the two, you get an almost unpredictable mix. Some moments there will be excess generation; other times there's a net deficit, and you'll need to draw power from the grid.

Homeowners with a wind turbine under these conditions will want to minimize their excess generation. Why sell valuable electricity to the utility for 3 cents per kilowatt-hour, when the utility will turn right around and sell it back to you for 10 cents per kilowatt-hour or more? To avoid selling a surplus, consumers have several choices: They can adjust their consumption by using dump loads as much as possible, they can match their consumption to wind availability as much as possible, or they can use wind turbines smaller than they might otherwise select.

It's a Catch-22 situation. Wind turbine cost-effectiveness increases with size. But to minimize the amount of energy you sell to the utility at a deep discount during times of surplus, you'll want to buy a wind turbine that produces only a portion of your own consumption. Under German wind conditions and electrical consumption, for example, only one-third of the electricity from a wind turbine sized to meet the annual domestic needs of a home will actually be used in the home. Two-thirds will be sold back to the utility. To use two-thirds of the consumption in the home and only sell one-third back to the utility, you need to use a wind turbine that produces half or less of your

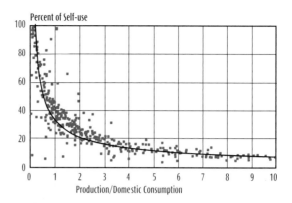

Figure 6-5. Degree of Self-Use. The percentage of on-site demand supplied by a wind turbine as a function of the wind turbine's annual generation and total consumption. This chart is derived from German data on more than four hundred household-size wind turbines—the most extensive in the world. There is some scatter in the data because some owners wisely shift discretionary consumption to periods when the wind is strongest.

annual domestic consumption. The situation is only slightly better with net metering, which also limits the size of wind turbine you can effectively use.

Net Metering

If the wind turbine is "behind the meter," as the utility would say—that is, serving on-site loads before any electricity reaches the grid—you would like to run your kilowatt-hour meter backward, selling any excess energy at the retail rate and buying what you need when you need it. In this way, you use the utility as a battery. The utility stores your energy until you need it. This is the essence of net metering.

To the utility, net metering means losing

---

5. DSIRE (Database of State Incentives for Renewables and Efficiency) is a comprehensive source of information on state, local, utility, and federal incentives in the United States that promote renewable energy and energy efficiency. See www.dsireusa.org.

a customer, and no business likes losing a customer. Consequently, they vigorously fight such policies. North American renewable energy advocates have spent years lobbying to overcome utility opposition to net-metering policies.

Where net metering is permitted, most utilities will replace your kilowatt-hour meter with

Breaking Free from Net Metering

For more than two decades North American renewable energy advocates have pushed net metering—the ability to run your kilowatt-hour meter backward—as the principal means to develop the distributed generation of renewable energy.

Net metering served a useful purpose in the dark days of the post–Jimmy Carter era. Net metering then was a call to arms for hobbyists and guerrilla solar activists out to prove a point—wind and solar works, your meter will run backward, and the lights will stay on.

But net metering was never intended to be a policy for the rapid development of the massive amounts of renewable energy that North America desperately needs. It could not do that alone. Retail electricity prices in North America, especially in much of Canada and in the Pacific Northwest, are abysmally—some would argue immorally—low.

Why? Because in most cases the price offset is insufficient alone to drive profitable renewable development. Subsidies are needed, and subsidy programs have had a checkered history in North America. Most subsidy programs have led to widespread abuses that have hurt renewables over the long term. Even today few subsidy programs require metering actual generation—one of the fundamental means for monitoring the success or failure of renewable energy programs. Worse,

subsidies are dependent upon the appropriation process, which is subject to political whim and competing needs for public funds to address the crisis *du jour.*

On top of that, there's typically a low limit on the amount of renewables that can be installed in net metering programs. If we really wanted renewables, why set a limit at all? Caps on program size serve only to protect utility markets. Utilities tolerate net-metering programs with low caps because the programs pose no serious threat to their markets. Thus, they—and the politicians who listen to them—limit the role renewables can play under net metering. But just to make sure, there's nearly always a limit on the size of any individual installation that can be installed under net metering programs, often the equivalent of one wind turbine. We certainly wouldn't want renewable energy to rock the utilities' boat.

Utilities, though, have little to fear. Net metering is self-limiting. Only those whose consumption can absorb all the production from a large turbine will choose to net meter. Customers such as Iowa's Schaefer Systems and Spirit Lake School will opt for larger turbines, because it makes economic sense to offset as much of their load as possible. Those for whom a 7-kilowatt turbine is a closer match to their needs will choose a 7-kilowatt machine. Artificial limits are unnecessary.

two ratcheted meters: one to register your purchases, and the other to register your sales to the utility. The utility then balances the account on a regular basis.

In true net-metering programs, the account is balanced annually. Beware: There are some false "net-metering" programs where the accounts are balanced monthly, and some

Net metering appeals to policy wonks because it rarely threatens entrenched electric utilities, and it gives politicians—as well as the advocates who promote these policies—the perfect cover for appearing to take action on renewable energy—while doing little of substance.

In the end, though, net metering won't get us where we want to be: massive amounts of renewables in the ground quickly. Net metering will never give us plus-energy houses or plus-energy buildings, because we often literally have to give our surplus electricity to the utility company for free. And where we don't, the payment for our excess generation is insufficient to justify the investment.

Europeans roll their eyes when North Americans speak of net metering. *"Was ist das?"* or *"Qu'est-ce que c'est?"* they can be heard saying. They don't bother with net metering.

How can Europeans be so successful if they don't use net metering? How can they have installed so much generating capacity that the Danes produced one-quarter of their electricity with wind turbines, and the Germans produced 15 percent of their supply from renewables in 2008?

The answer is surprisingly simple: They pay for it. Europeans set a tariff or a payment per kilowatt-hour for each renewable energy technology, one sufficient to cover the cost of generation plus a reasonable profit. This ensures that they get the kind of renewables they want, where they want, at the pace they want. The results speak for themselves. No taxpayer subsides are needed. Electricity consumers pay for renewable energy development by paying a charge on their electric bill just like the charge they are paying now for fossil-fired generation. Taxpayers don't pay. Those who consume more electricity pay more. Those who use less electricity pay less.

Net metering was always just a stopgap—a policy that could buy us time until the political climate changed and we could implement serious renewable energy policies. That time has come.

The time for half measures—for timid responses like net metering—is past. The public, and now some progressive politicians as well, are demanding more aggressive policies. For distributed renewable energy to make the substantial market inroads needed in the huge North American market, advocates need to break free from the net-metering straitjacket.

We need new policies, such as electricity feed laws, that have a proven record of results. For more on these policies and how they can benefit renewable energy development in North America, visit www.wind-works.org and go to the feed law pages. — PAUL GIPE where generation fed into the grid is paid substantially less than the retail rate. (Yes, some utilities call this "net metering.")

There are often onerous restrictions on net metering, who can use it, the size of the system permitted, and the size of the total program. Most jurisdictions limit the size of the wind turbine that can be used under net metering. Some programs limit turbines to as little as 10 kW; others allow as much as 1 to 2 MW. Effectively, net-metering programs limit you to one wind turbine sized to offset your load.

Most states and provinces limit the total amount of generating capacity that can be installed under net-metering programs. The limits are often meager, typically just a small percentage—sometimes just a fractional percentage—of the utility's total load.

In most states net-metering regulations affect only investor-owned or regulated utilities, thus excluding the many consumers connected to rural electric cooperatives or municipal utilities. Only a few states offer net metering to all electricity consumers.

Net metering can make interconnecting a small wind turbine with the utility more financially attractive than it might otherwise be. With net metering, you can use a larger, more cost-effective wind turbine than you would where net metering was prohibited, but you are still limited. If you choose to use a larger wind turbine because it's more cost-effective than a smaller one, you may produce more electricity than you consume. Under most net-metering programs, utilities will pay only a token amount for your surplus generation—and in some states they won't pay anything at all. It's tough to make a good economic case for wind energy under these conditions.

European Distributed Generation

The small turbine industry in Denmark began much the same way as that in North America. The first turbine using an induction generator was connected to the grid without permission, and unbeknownst to the utility, in the mid-1970s. This could have been the world's first example of "guerrilla wind," the counterpart to the 1990s "guerrilla solar" movement countenanced nearly two decades later by *Home Power* magazine. But the Danish industry has since followed a far different trajectory than North America's. The once small turbines of Danish manufacturers gradually grew bigger, becoming the core of the world's commercial wind power industry.

Early Danish turbines were about the same size as those being installed in the United States at the time: 7 kW, 10 kW, and 15 kW. But they and their manufacturers grew because there was public and political support for them to do so. This support manifested itself in flexible siting policies and in interconnection policies that went well beyond net metering. First Denmark, then later Germany implemented policies that gave wind turbines priority access to the grid. Equally as important, Denmark and Germany paid a fair price—a price sufficient to make projects profitable—for wind-generated electricity.

Instead of limiting the interconnection of a wind turbine to a consumer's own property for directly offsetting consumption (as in net metering), Denmark permitted the wind turbine to be located anywhere within the township, and eventually within the neighboring township as well. Thus there was no meter to run backward.

Figure 6-6. Lynetten Urban Cooperative. The Lynetten cooperative owns four of the seven 600 kW wind turbines on a breakwater in Copenhagen's commercial harbor. The turbines, in operation since 1996, are visible from the Danish parliament building, from the famous *Little Mermaid,* and, as shown here, from the Danish historical park Trekroner. Lynetten is just one of three wind projects within urban Copenhagen. Another is the famous 40 MW Middelgrundens offshore project, also cooperatively owned. For more information on Lynetten, visit http://lynettenvind.dk; to learn more about Middelgrunden, see www.middelgrunden.dk.

There would be only one meter, a new meter where none had existed before, and it would register delivery—and sales—to the utility.

Danish politicians were responding to a widespread demand for the ability of the public to develop their own renewable resources, and especially to install their own wind turbines. But like politicians everywhere, they were cautious. They didn't want to open the floodgates. Denmark, with a long history of cooperative rural economic development, targeted attractive wind tariffs and other incentives at local ownership. Each homeowner was allowed to buy shares in the generation from a local wind turbine equivalent to 1.5 times domestic consumption. Eventually, the limit was raised so that a family would not only offset their electricity consumption, but also the equivalent of their total energy consumption.

Denmark thus created a program that not only permitted interconnection with the grid but in fact encouraged the public's participation in doing so. By the 1990s 5 percent of the Danish population owned shares in a neighborhood wind turbine. Today, these cooperatively owned wind turbines can be seen distributed across the Danish landscape from the far northwest of Jutland to the city of Copenhagen (see figure 6-6, Lynetten Urban Cooperative).

Similarly, in 1991 Germany responded to

public clamor for the ability to generate renewable energy by launching its first electricity feed law, the *Stromeinspeisungsgesetz:* literally, the law on feeding electricity into the grid. As its name says explicitly, this law permits connection to the grid, and the delivery or "feeding" of electricity into the grid. Germany's feed law

Distributed Generation in North America

Until recently, most commercial wind development in North America has been the construction of large, central-station wind power plants. In the early 1990s a large project could be 50 to 80 MW. Today projects of 200 MW to 300 MW

The German feed-in law went beyond the Danish program by placing no restrictions on how much electricity could be generated by each owner. There is no mention of net metering or offsetting one's own consumption.

---

required utilities to not just permit but also enable connections with renewable energy generators, and it spelled out what utilities were to pay for this generation.

The German feed-in law went beyond the Danish program by placing no restrictions on how much electricity could be generated by each owner. There is no mention of net metering or offsetting one's own consumption.

Danish renewable energy policy specifically encouraged distributed local ownership. German policy did not. Distributed development in Germany largely resulted from the complexity of siting wind turbines in a densely populated country.

But in both Denmark and Germany, it was in everyone's financial interest to install the size wind turbine that made the most economic sense at the time. Farmers, homeowners, and wind developers were not forced to choose a wind turbine smaller than they needed because they could only offset their own consumption.

are routinely announced. (For example, the Cape Wind project offshore from Nantucket could be more than 400 MW in size.) These projects are all connected at transmission voltages. They are far too large to be added to local distribution lines.

However, many wind projects in Denmark, Germany, France, and the Netherlands are connected to local distribution lines. The size of wind project, or the number of turbines it can support, varies widely depending upon the size of the conductors (the lines carrying the electricity), other generators already on the line, and many other factors. Projects could range from 5 MW (2–3 turbines) to 20 MW (the equivalent of 10 or more turbines). See table 6-1, Distribution Voltage and Wind Capacity.

Projects of this scale, the "third way" of developing wind energy, can be placed close to the loads they will serve: towns, cities, scattered farmsteads, industrial zones, and so on.

The upper Midwest has seen more distributed wind development than any other region of North America, and Minnesota in particular

Table 6-1: Distribution Voltage & Wind Capacity

Voltage kV	Capacity MW
25	6–10
45	15–20

Table 6-2. Minnesota Distributed Wind

	MW	Percentage*	Units
Small Developer	104	12%	93
Farmer Owned	74	8%	56
Locally Owned	72	8%	61
Municipal Utility	19	2%	18
Rural Electric Cooperative	6	1%	9
College/University	5	1%	3
School	1	0%	1
	281	32%	241

* Percentage of 902 MW

Source: Windustry.org, March 2007

more than 2 MW, but as turbines grew ever larger this limit was eventually lifted.

Now there are single large wind turbines or small clusters of turbines throughout the windy parts of the state, and even some in less windy areas. Large wind turbines can be found at schools and universities, on farms, and in privately owned, mini wind farms. For example, there's one at Carleton College in Northfield, Minnesota and across town another at St. Olaf College. All told, distributed wind turbines accounted for almost one-third of the wind generating capacity in Minnesota in 2007 (see table 6-2, Minnesota Distributed Wind).

The future promises a rapidly growing contribution from distributed wind in North America as the wind industry, community and renewable energy activists, and political leaders all realize its economic and energy benefits. While large-scale wind farm development will continue to grow rapidly, and some individuals will continue to install small wind turbines, the prospect of locally owned, distributed wind will become increasingly appealing.

stands out. For more than a decade Minnesota has encouraged distributed wind development alongside large commercial projects. The state first limited distributed wind projects to no

Community Wind

7

Until recently, there have been two principal ways of developing wind energy in North America. The first and most prominent has been commercial developers installing giant wind farms for delivering bulk electricity to the grid. The second has been homeowners and farmers installing small wind turbines for their own personal benefit. There is, however, a third way to develop wind energy: community wind.

Community wind is wind energy for the rest of us. It's wind energy that everyone can participate in, regardless of where they live—whether in a city high-rise or in a small town. You don't have to live on a windy plot of land in a rural area to invest in wind energy—to make a difference in moving North America toward its renewable energy future.

Much like investing in a mutual fund, participating in community wind can spread the cost—and risk—of developing wind energy over a large number of shareholders. Instead of installing a small wind turbine in your backyard, the cost and maintenance of which are entirely your responsibility, you can join with your neighbors and together invest in a commercial-scale wind turbine in your community or nearby.

Community wind is part of North America's growing community power movement. Grassroots groups are campaigning for policies that make community power possible from Oregon to British Columbia, from Massachusetts to New Brunswick, from the upper Midwest to Ontario and Manitoba.[6] They are trying to replicate in North America the success of local ownership in powering the rapid growth of renewable energy in continental Europe.

What Is Community Wind?

There's no hard-and-fast definition of *community wind*. This in part results from our attempt to describe in English a uniquely continental European phenomenon found notably in Denmark, Germany, and the Netherlands. We typically think of share ownership structures that allow local investment in renewable energy projects. But farmers and small businesses individually also own a large portion of the wind and biogas projects in these countries,

Figure 7-1. Joint Ownership. Two of five 600 kW NEG-Micon turbines along a drainage canal in the Wieringemeer polder of Nord Holland. Farmers on either side of the canal own two of the turbines, one is owned by the local utility, and the manufacturer owns one.

6. See Greg Pahl's book *The Citizen-Powered Energy Handbook* at www.chelseagreen.com for more on the community power movement.

Unlike net-metering policies in North America, European wind energy policies don't discriminate on the basis of size. Community groups can buy the most cost-effective wind turbine for their site, regardless of size.

and in some cases local ownership is mixed with corporate ownership (see figure 7-1, Joint ownership).

Similarly, the term *community wind* often implies small groups jointly owning large wind turbines, but this isn't always the case. While many community power projects fall into the category of distributed generation, some "community wind" projects are quite large and fall into the category of wind power plants connected at transmission voltages, albeit owned by those nearby or within the neighboring region.

Community renewables may include small solar power systems operated by individual farmers and homeowners, but typically do not include small household-size wind turbines. Wind turbines are significantly more cost-effective on a commercial scale, and the community wind movement in continental Europe has focused exclusively on commercial-scale turbines.

Unlike net-metering policies in North America, European wind energy policies don't discriminate on the basis of size. Community groups can buy the most cost-effective wind turbine for their site, regardless of size. Though older community-owned wind turbines in Europe, especially in Denmark, appear small by today's standards, they were large turbines in their day.

Denmark's *Fællesmølle*

Denmark has long been known for the high concentration of jointly owned wind turbines (*fællesmølle*), as well as commercial-scale wind turbines owned individually by farmers. Individuals and families collectively own

Figure 7-2. Modern Danish Wind Cooperative. Four of the eight turbines in this project "offshore" in Denmark's Limfjord are owned by the Thyborøn-Harboøre Vindmøllelaug, or wind turbine cooperative. The Vestas V80s were installed in 2003. For an aerial view of the site and a gallery of installation photos, visit www.roenland.dk.

one-quarter of the 3,200 MW of wind turbines operating in the country. Individual farmers and co-owned installations represent nearly two-thirds of the installed capacity, representing about 2,000 MW. Only one-tenth of Denmark's wind generating capacity is owned by traditional electric utilities (see figure 7-2, Modern Danish wind cooperative).

Table 7-1: Co-Op and Farmer-Owned Wind Turbines in Europe			
	Farmer	**Co-op**	**Corporate**
The Netherlands	60%	5%	35%
Germany	10%	40%	50%
Denmark*	64%	24%	12%
Spain	0%	0%	100%
Great Britain	1%	1%	98%
Source: Dave Toke, University of Birmingham, 2005, updated to Toke 2008. *Onshore.			

Germany's *Bürger* Wind Movement

Similarly, the *Bürger* or citizen movement in Germany forms limited liability companies that seek share investments from local landowners and neighboring communities. These *Bürgerbeteiligungen,* or citizen-owned companies, build community wind, solar, and biogas plants. Typically, the project developers raise as much equity as possible from the local community, and then expand their net to the region. Finally, if there's not enough capital to build the project, they open investment up to the entire country.

Nearly 40 percent of the 22,000 MW of German wind capacity at the end of 2007, or 8,000 MW, is *Bürger*-owned, while another 10 percent, or 2,000 MW, is owned directly by farmers. In contrast, less than 2 percent of the wind generating capacity in the United States is locally owned, and nearly all of that is in one state, Minnesota (see table 7-1, Co-op and Farmer-Owned Wind Turbines in Europe).

At the end of 2007, Germany had more installed wind generating capacity than the entire North American continent (nearly 17,000 MW in the United States, less than 2,000 MW in Canada, and less than 100 MW in Mexico).

Traditional corporations in various forms have installed one-half of the German wind capacity. Yet landowners and small investors have developed the other half, raising and managing investments worth nearly $20 billion—and that's just for wind energy!

German homeowners and farmers have likewise installed 90 percent of the 4,000 MW of solar photovoltaic systems in the country—twice the amount installed in Japan, which was once the world's leader in solar cells—and four times the amount installed in North America. The same is true of biomass and biogas. The citizen-owned renewable energy movement is a critical factor in the success of renewable energy in Germany because it enables every citizen to participate if he or she chooses to do so.

The Danish and German shared-ownership model has now been exported to France as well as to North America.

Size of Community Wind Projects

Community wind projects in Europe and in North America can be of any size. The smallest projects entail one turbine like WindShare's single Lagerwey on Toronto's harbor front (see figure 7-3, Toronto's WindShare). Others can be quite large, as the following examples illustrate.

- The Elkhorn Ridge Wind project will be 80 MW when completed. It is expected to comply with Nebraska's Rural Community-Based Energy Development Act and thus qualify as community wind by Nebraska's definition.
- The Paderborn Bürgerbeteiligung (citizen-owned wind project) comprises several separate projects totaling 111 MW.
- Bear Mountain Wind, initiated by Peace Energy Cooperative in British Columbia, will be 120 MW when completed in 2009. While this project was initiated by the cooperative, a private developer is leading the financing and construction.
- The Samsø Offshore wind plant comprises 10 turbines of 2.3 MW each, for a total capacity of 23 MW. Two of the turbines are owned locally by residents of Denmark's "Renewable Energy Island." The rest are owned by outside investors.
- Middelgrunden, the famous 40 MW project offshore from Denmark's capital city of Copenhagen, was developed by and is 50 percent owned by Middelgrundens Vindmøllelaug. The other 50 percent is owned by the city's municipal utility. Middelgrunden is the largest of Denmark's cooperatively developed wind projects.
- Though not yet built, the development cooperative Offshore-Bürger-Windpark Butendiek plans to build a 200 MW project off the west coast of the Jutland Peninsula just south of the Danish border. Clearly the proponents of this

Figure 7-3. Toronto's WindShare. The Toronto Renewable Energy Cooperative developed WindShare, the first wind turbine cooperative in Canada. WindShare's turbine in downtown Toronto has become a landmark. Like similar projects in Germany and Denmark, the turbine is co-owned by the cooperative and the local municipal utility, Toronto Hydro. There are more than 400 members, including the lead singer for the popular Canadian rock band Barenaked Ladies, Steven Page. For more information, visit www.windshare.ca.

Figure 7-4. Velling Mærsk-Tændpibe. When installed in the mid-1980s, Velling Mærsk-Tændpibe was the largest wind plant in Europe. It consisted of 100 turbines of various sizes co-located on a polder south of Ringkøbing on the west coast of Denmark's Jutland Peninsula. One-third of the turbines are cooperatively owned.

project, farmers from northwest Germany, many of whom had already successfully developed their own local community wind projects, don't feel that size in and of itself is a limiting factor.

• When installed in the mid-1980s, Velling Mærsk-Tændpibe was the largest wind plant in Europe. It consisted of 100 turbines of various sizes, co-located on a polder south of Ringkøbing on the west coast of the Jutland Peninsula. One-third of the turbines are cooperatively owned. Though the turbines were state-of-the-art then, they are small by today's standards. The total project amounted to only 13 MW, and the cooperative portion was only 2.6 MW. Still, for the co-op investors and nearby communities this was a big project, and it remains visually impressive even today (see figure 7-4, Velling Mærsk-Tændpibe).[7]

7. The Tændpibe cooperative was sold in the mid-2000s, earning investors an annual average rate of return of 14 percent over the two decades they owned the project.

German Feed-in Tariff Example

Community wind is possible in Germany because feed-in tariffs allow profitable development in many areas of the country, not only the windiest sites. Thus farmers can install large wind turbines on their own land profitably, or groups of farmers can do so together.

Table 7-2 is a small extract of the feed-in tariffs for renewable energy in Germany. A full list of the renewables tariffs in Germany as well as in many other countries, such as France, Spain and Switzerland, can be found at www.wind-works. org/FeedLaws/TableofRenewableTariffsorFeed-InTariffsWorldwide.html.

The German system pays all wind turbines the same price or tariff for the first five years of operation. After the fifth year, the productivity of the turbines is determined and compared with the "reference yield." If the productivity is more than the reference yield, the payment is reduced; if it is less than the reference yield, the premium payment is extended. For example, if the productivity of wind turbines on land is 60 percent of the reference yield, the owner will be paid €0.092/kWh for the entire 20 years. If the productivity is 150 percent or greater than the reference yield, the owner is paid €0.052/kWh for years 6 through 20. In this way, Germany ensures a "fair" profit but limits excessive or "unfair" profits to protect consumers.

Ownership Requirements

While community wind is not overtly prohibited in North America, there are very few opportunities for it to flourish. (Organizations such as Windustry in Minnesota and OSEA in Ontario hope to change that.) There are few incentives specifically targeted at community wind. And some renewable energy incentives, such as the production tax credits in the United States, were intended for projects developed by large, traditional corporations and consequently are of little use to community wind projects that have no or limited tax liability.

Like Denmark's early programs, some jurisdictions provide incentives specifically for community wind, and restrict participation to those that meet specific criteria. Minnesota and Nebraska define in statute what qualifies as community wind; New Brunswick is weighing a similar policy.

It's noteworthy that in Germany there are no ownership requirements to qualify for the feed-in tariffs that make German-style community wind possible. Projects developed by traditional corporations qualify for Germany's attractive feed-in payments. When the Ontario Sustainable Energy Association (OSEA) chose to replicate the German model of community wind development in North America, it chose a similarly open policy.

OSEA, a spin-off from the cooperative that developed Toronto's WindShare turbine, is the leading nongovernmental organization in Canada promoting community renewable development. OSEA realized early on that existing policies in Ontario favored traditional corporate development, and in order to move

Table 7-2: Advanced Renewable Tariffs in Germany				
Erneuerbare-Energien-Gesetz (EEG) and EEG 2009 Tariffs				
	Years	Tariff €/kWh	1.63 ~CAD/kWh	1.30 ~USD/kWh
Wind*				
On Land				
60% Reference Yield	20	0.092	0.15	0.12
100% Reference Yield	12.4	0.092	0.15	0.12
150% Reference Yield	5	0.092	0.15	0.12
All	To Year 20	0.050	0.08	0.07
Offshore		0.15	0.24	0.19
60% Reference Yield	20	0.15	0.24	0.19
100% Reference Yield	16	0.15	0.24	0.19
150% Reference Yiled	5	0.15	0.24	0.19
All	To Year 20	0.035	0.06	0.05
Solar Photovoltaic				
Freestanding	20	0.319	0.52	0.41
<30 kW rooftop	20	0.430	0.70	0.56
<100 kW rooftop	20	0.409	0.67	0.53
>100 kW rooftop	20	0.396	0.65	0.51
>1,000 kW rooftop	20	0.330	0.54	0.43

*There is a continuum between reference yields for wind energy that's not explicitly stated in the table. See the English language version of the EEG for an explanation.

beyond one community-owned turbine, new policies needed to be put in place.

In 2004 OSEA launched a campaign to bring electricity feed laws, like those used so successfully in Germany, to Ontario. At the time, OSEA deliberately chose not to place any ownership requirement on the program. It was the association's intent that the program be open to all participants: farmers, homeowners, cooperatives, indigenous communities, and traditional corporations. OSEA chose this route because it could not envision all the possible combinations of corporate ownership, individual ownership, and mixed-ownership models that would serve the needs of both community economic development and the rapid growth of renewable energy.

Others have chosen different models. To qualify for the Community-Based Energy Development (C-BED) tariffs in Minnesota, 51 percent or more of the equity in a project must be owned by Minnesotans. Moreover, no single entity may own more than 15 percent. Minnesota wanted to spread the benefits of the C-BED program to as many participants as possible, rather than see development concentrated in the hands of a small number of wealthy individuals. An exception is made for small projects. For practical reasons, the Minnesota law allows a single individual to own up to two wind turbines.

Nebraska has followed a path similar to Minnesota's. Nebraska's Rural Community-Based Energy Development Act requires that 33 percent of the payments from a wind project over the 20-year contract period must go to Nebraskans if the project is to qualify. For example, the $140 million Elkhorn Ridge wind plant near Bloomfield, Nebraska, is being built in partnership with Midwestern Wind Energy, a Chicago developer. The project will be the largest in Nebraska when completed by the end of 2008, and one-third of its financial benefits will flow directly to Nebraska residents.

Elsewhere in North America, there are no specific criteria on what is not community wind. At one end of the spectrum is the WindShare cooperative. It owns half of the wind turbine at Toronto's Exhibition Place. The other half is owned by Toronto Hydro, the municipal utility. Like projects in Denmark, the cooperative led development, and like the *fællesmølle* cooperatives in Denmark, WindShare is 100 percent owned and controlled by investors living in the region. Each shareholder has as much say as another in the management of

Figure 7-5. Local Democracy in Action. Basia Pioro collects votes from WindShare members at the 2006 annual general membership meeting in Toronto. Each member has one vote regardless of the number of shares he or she owns. WindShare also sponsors an annual summer picnic beneath the wind turbine. Dairy coops have been a successful business and community development model in both Denmark and North America. Wind coops could be equally successful.

In Denmark and Germany, wind energy was developed by rebels, those who wanted to break with conventional ways of generating electricity. In German they're called stromrebellen (electricity rebels), and we can thank them for the modern wind and solar industry.

their wind turbine. WindShare has become a North American model for engaging the public's imagination and active participation in developing renewable energy (see figure 7-5, Local democracy in action).

At the other end of the spectrum is Peace Energy Cooperative's Bear Mountain Wind project. When commissioned in 2009, the project will represent a total investment of nearly a quarter of a billion dollars and will be among the largest wind projects in Canada. The 250-member Peace Energy Cooperative initiated the project by obtaining leases on public land and measuring the wind resource themselves.

The cooperative then formed Bear Mountain Wind in partnership with a private developer, and hopes to raise $5 to $12 million for its portion of the project. Depending on the proportions of debt to equity for installation of 60 Enercon E-82 turbines, Peace Energy's share could represent from 9 percent to 25 percent of the total equity in the project. Unlike WindShare, Peace Energy Cooperative will have neither a controlling interest in nor managerial control of the project.

Farmers on the south shore of Quebec's Lac St-Jean have tried a similar approach. The farmers formed a cooperative, Val-Éo, and a limited partnership to develop their wind resource. (The cooperative then acts as the general partner of the limited partnership.) The farmers provided all the upfront risk capital for the first phase of developing a 50 MW project. Like Peace Energy Cooperative in British Columbia, Val-Éo chose to work with a private partner to eventually build and operate the wind plant.

In the Val-Éo model, the cooperative can provide up to 50 percent of its own equity in the project. In an unusual twist, however, Val-Éo will maintain 50 percent of the voting shares regardless of the percentage of equity actually owned by community members once the project is built.

Unfortunately, Quebec, like many other provinces and states in North America, has not been supportive of community wind. While not openly hostile, these jurisdictions have chosen a different development model that effectively bars community wind by transferring the right to develop wind energy to a few chosen corporations.

Nowhere is it easy to develop community wind, but it is particularly difficult in North America. It's not easy in Denmark and Germany, either. You have to be a rebel to develop community wind. A streak of stubbornness helps, too.

Electricity Rebels

In Denmark and Germany, wind energy was developed by rebels, those who wanted to break with conventional ways of generating

Figure 7-6. Fesa *Bürgerbeteiligung.* **Two Enercon E-70s on the flanks of Schauinsland in southern Germany's Black Forest. The 98-meter (320 ft) concrete towers used here raise the 1.8 MW turbines well above the treetops. Developed by Josef Pesch's Fesa, these turbines are part of a group of six in Regiowind Freiburg. Investors live in the "solar city" of Freiburg on the valley floor in the background. Regiowind Freiburg produces 2 percent of the city's electricity.**

electricity. In German they're called *stromrebellen* (electricity rebels), and we can thank them for the modern wind and solar industry. One is Josef Pesch. He was named a *Strom Rebel* by *Elektrizitätswerke Schönau* (EWS) in 2005. The good folks at EWS know electricity rebels when they see them. They themselves were dubbed *Stromrebellen* by popular German author and lecturer Franz Alt for leading a successful referendum that gave control of the local distribution company to the villagers of Schönau.[8]

Pesch is the man behind FESA, a private firm that develops community-owned renewable energy projects. FESA organizes these *Bürgerbeteiligungen* (share cooperatives) around solar, wind, and wood-pellet-fired district heating plants in southern Germany's famed Black Forest.

"Farms here are not very rich, because they are so small," says Pesch in explaining why southern Germany's farmers were quick to consider renewables as a new source of income. "Farmers were the pioneers of wind energy in Germany." And despite their low incomes,

8. To read "The Electricity Rebels of Schönau: The Village That Built Their Own Solar Utility," see www.wind-works.org/articles/SchoenauStromRebels.html.

they have fewer problems finding money than others, because farmers have land they can use as collateral, Pesch explains.

"We're creating a new class of entrepreneurs," he adds, by engaging the public in developing renewable energy. "Wind would not exist today if it relied only on the investments of the big utilities. The earnings [from wind energy] are too low because the winds are too low [here in Germany]. Wind would never have been developed"—because the returns on investments are not attractive enough for utility companies.

Yet private citizens (the Bürgers of Bürger wind projects) are willing to invest for such earnings. Half of the investors in FESA's Freiamt project, for example, probably never invested in anything before, Pesch suspects. Now they are co-owners of a small wind farm.

In another FESA project, Regiowind Freiburg, the 521 owners invested $6 million in a $20 million project comprising six turbines. Half to two-thirds of the investors were local, another 10 percent were from the state of Baden-Württemberg, and the rest came from elsewhere in Germany (see figure 7-6, FESA *Bürgerbeteiligung*).

In this manner, small local projects have been developed all across Germany.

In the Bürger Windpark movement, says Henning Holst, the citizens or Bürgers provide the equity in a limited liability company. Typically, the developers raise as much equity from the local community as possible. If there's not enough capital to build the project, they open investment up to larger and larger areas.

Holst develops citizen-owned wind projects in Germany's northernmost state, Schleswig-Holstein. His engineering office is located in Husum, the center of the German wind universe and the biennial wind extravaganza, Husum Wind, the world's largest exhibition of wind turbines.

One of Holst's most significant projects was repowering a portion of the wind plants on the island of Fehmarn off Germany's North Sea coast. His task was to upgrade 46 MW of older wind turbines with fewer larger turbines and, in doing so, substantially raise capacity to 136 MW. He estimates the project will cost nearly $200 million when completed.

Unlike smaller projects connected at the distribution voltages elsewhere in Germany, the Fehmarn turbines are grouped in clusters and connected at transmission voltages. Holst had to maintain the existing connection to the grid while adding the new generation via a 31 km, 110-kilovolt subsea cable to the German mainland.

While not an example of distributed generation, Holst's Fehmarn repowering is an example of how small entrepreneurial companies combined with the Bürger wind movement have built major projects. Size is not necessarily a limiting factor.

Another *Stromrebel* in Germany's far north is Wolfgang Paulsen, one of the owners of a Bürger wind project near the Danish border. Paulsen lives in Bohmstedt, a village with a grand total of 650 people, 20 farms, and, as Paulsen likes to emphasize, one pub—an essential element of German life.

In 1998 Bohmstedt's residents installed nine 600 kW turbines. The turbines produce 11 million kilowatt-hours per year, enough electricity for 3,000 north German households—far more than that consumed in Bohmstedt itself.

"Our objective was to involve as many citizens as possible," says Paulsen. Each participating family invested risk capital of about $2,500 each. Together they raised $75,000 for the first phase: planning and development. If the project didn't move forward, their initial investment would be lost. Even in the favorable renewable energy climate of Germany, investment in community wind requires a willingness to take risks. This is an example of what FESA's Pesch describes as a new class of German entrepreneurs—farmers and villagers.

Why Community Wind?

As explained in the introduction to this chapter, community wind enables everyone to participate in the renewable energy revolution, not just corporate CEOs or those fortunate enough to live on a windy plot of land. For wind energy—and renewables in general—to reach their full potential, they must be accepted by the communities of which they are a part.

Large conventional power plants, because there are so few of them, can be built over

---

As Dutch farmers are quick to say, "Your own pigs don't stink."

They then needed to raise 15 percent of the $7.5 million cost of the project. The remainder was financed with bank loans. Ultimately, each family invested a minimum of $35,000 per share.

In explaining the philosophy behind the Bürger wind movement in northern Germany, Paulsen is frank. "Wind is a local resource. It is our resource. We want renewable energy, and we want to make money out of it. So we do it ourselves."

FESA's Pesch has similar views. "Those who consume electricity should also be part of the solution," he says. They should develop renewable energy themselves and not leave it to the utilities or multinational companies. "This is the right thing to do."

Josef Pesch, Henning Holst, and Wolfgang Paulsen are pioneers in a different way of developing and using wind energy. They are indeed *Stromrebels*.

the objections of local residents. Not so with distributed renewables. Distributed renewables are different because there must be thousands, possibly millions, of wind turbines, solar panels, and biogas plants scattered across the vast breadth of North America to meet the energy needs of its inhabitants. To succeed, distributed renewables in some form or another must be welcomed into nearly every community.

As Dutch farmers are quick to say, "Your own pigs don't stink." If members of the community have a stake in a group of wind turbines, or own money-generating solar systems on their rooftops, they are less prone to suggest putting them somewhere else.

In community wind, less of the revenue leaves the community for distant financial centers, such as Toronto, New York, or Chicago, than in traditional corporate development. Community wind puts the revenue—the profit—from renewable generation into the

Table 7-3: Community Wind Economic Impact			
Annual Direct, Indirect, and Induced Effect on Value Added			
	Community Wind	Community Wind	Corporate Wind
	5%	8%	
Value Added	$ 1,300,000	$ 640,000	$ 250,000
Jobs	14.5	8.2	4.3
Source: Arne Kildegaard, University of Minnesota, Morris			

Community Wind North American Sources of Information

- Ontario Sustainable Energy Association: www.ontario-sea.org
- Windustry: www.windustry.org
- Wind-Works.org: www.wind-works.org/articles/community.html
- *The Citizen-Powered Energy Handbook* by Greg Pahl: www.chelseagreen.com

pockets of the communities where the wind turbines will be installed. More of the earnings stay in the community, circulating among residents and shopkeepers, creating more local jobs than otherwise (see table 7-3, Community Wind Economic Impact).

Here's another way to look at it. Consider the example of one 80-meter (260-ft) diameter wind turbine at a modestly windy site with an average wind speed of 6 m/s (13.4 mph). Let's say the 2 MW turbine generates 4 million kilowatt-hours per year, and let's assume we can sell that electricity for $0.10/kWh. This one lone wind turbine will earn gross revenues of $400,000 per year. If a farmer leased his land to an outside wind company he might, at best, receive a royalty of 5 percent, or $20,000 per year. This is a new cash crop for the farmer that he didn't have before. It's far better to earn 5 percent than just to curse the wind.

Now let's assume that the farmer is willing to take the risk of managing his wind investment. And let's assume that the total cost of the project, after accounting for all operating expenses, can be recovered within 10 years. During the second 10 years of the 20-year life of the turbine, the farmer potentially could earn $4 million, less annual operating expenses.

Over the same 20 years, the farmer who leased his land would earn royalties of about $400,000. Certainly leasing land to a commercial wind developer is easier and less risky than making an investment in a 20-year, capital-intensive project like a wind turbine. And farming is certainly a risky business to begin with. On the other hand, most wind leases in North America pay only a small fraction of the 5 percent royalty common in Europe. Many pay only half that—some even less.

In this example of one commercial-scale wind turbine, community wind development could potentially pump millions of dollars into the local economy. The traditional land lease, in comparison, would deliver an order of magnitude less.

The demand for renewable energy in North America is so great that we will need a mix of corporate development of large wind farms, community wind, and, where suitable, individually owned wind turbines. We will need it all. Unfortunately, communities, farmers, and other rural landowners in most states and provinces have few options but to lease their land.

What's Required

Europeans have been successful in the rapid development of wind energy for two reasons. They have broad public support, in large part due to community participation. Communities and individuals (farmers as well as homeowners) have been able to participate because of electricity feed laws that not only allow their connection to the grid, but also specify the price that they will be paid for their generation.

These payments for renewable energy generation, or feed-in tariffs, are transparent and equitable. Everyone knows what is paid for each kilowatt-hour and for how long they will be paid. And everyone can receive these payments, regardless of their position in society, rich and poor alike. With information on the tariff that will be paid, any homeowner, farmer, or community group can calculate whether an investment in a wind turbine will be profitable—or not.

And just as important, every banker can make the same calculation as well. Bankers, too, can quickly determine whether an investment is a good risk or not. Where the payments are high enough, or the wind is strong enough, the banks know they can lend money to community wind projects—and expect to be paid back. With bank financing, a community wind project begins to look much like a traditional corporate wind project, except that more of the benefits stay in the local community.

Though only Ontario has implemented renewable energy tariffs to date, there is a growing movement across North America to adopt electricity feed laws like those in Europe. With well-designed feed-in tariffs, we could see a boom in not only small wind turbines connected to the grid, but in community wind as well.

Investing in Wind Energy

8

Wind turbines, large or small, are not cheap. Nor is buying, installing, operating, and maintaining a wind turbine simple or easy. But it can be worthwhile, in both economic and environmental terms. The challenge is finding the combination of factors that makes a venture into wind energy profitable.

When *Wind Energy Basics* was first published a decade ago, the book focused on how to buy and install micro and mini wind turbines. This edition, while continuing to include material on small wind turbines, broadens the scope to large wind turbines that may be owned by farmers or groups of people in a venture of shared ownership. For this reason, the subject of this chapter looks at ways to invest in wind energy other than simply buying a wind turbine.

For some, "buying" a wind turbine implies installing it and walking away, as when you buy a major appliance. You use it, but otherwise forget about it. Not so with a wind turbine, large or small. It becomes part of your life and requires your periodic attention. Sometimes it requires a lot of your attention.

Another way of saying that wind energy is not cheap is to describe it as capital-intensive. The fuel—the wind—may be free, but buying the machinery to harvest it is not. You have to pay nearly all of the cost of wind energy up front. To recover those costs and earn a profit, the wind turbine has to operate for a long period of time—say, two decades or more. Thus, the longevity of the turbine is critical to keeping the cost of wind energy down and the profits up.

It's important to note, says Ken Starcher of the Alternative Energy Institute at West Texas A&M University, that even though wind is not cheap, it can be both affordable and a good buy—over the long term.

Because wind turbines must be operated over the long term, cost so much up front, and demand our attention and care for such long periods, it's much better to view them as an investment than simply as an appliance. Rather than "buying" a wind turbine, we are "investing" our time, and our money, in wind energy. This is more than a philosophical difference. It governs not only how we design wind turbines and care for them, but also how we relate to them. As with marriage, it's important to get this relationship right, because if all goes well, you're going to be together for a long, long time.

Longevity

How long do wind turbines last? Some turbines operate for only a few years. Others are still running after more than a quarter century. After its installed cost and its productivity, a wind turbine's longevity is probably the next most critical factor in whether a wind turbine investment is profitable or not.

Turbines that last a long time and are less costly to maintain are worth more. Some old windchargers from the 1940s operated for decades on end with little maintenance. In the modern era of wind energy, there are thousands of wind turbines that have been in service nearly three decades and are still churning out the kilowatt-hours.

Some small wind turbines, notably lightweight machines, may survive for as little as five years before requiring replacement. Others, with regular maintenance and the occasional repair, can operate for decades. While some small wind turbines last for many years, many micro turbines don't do nearly as well.

Longevity of Micro Wind Turbines

Beech Mountain, near Boone, North Carolina, is famous as the host of one of the United States' early and ill-fated experiments with large wind turbines during the 1970s. It's in a class 5 wind regime in the American system with an average wind speed at its 1,560-meter (5,100 ft) summit of about 7 m/s (16 mph). Beech Mountain is a good place to put small wind turbines through their paces.

Appalachian State University operates a test site atop Beech Mountain, and ASU's Brent Summerville has tested several well-known small wind turbines. For example, he observed that there were no problems with an Air X, Southwest Windpower's popular micro turbine. It ran reliably for four years before failing.

Mick Sagrillo has a preference for "heavy metal"—that is, rugged, more massive turbines. Summerville's experience on Beech Mountain

also indicates that heavier machines appear more robust than lightweight turbines. "We like quiet and reliable, because they're the best to live with," says Summerville. And *reliability* is another word for robustness, the ability to withstand gusty winds for years on end.

These are relative measures, often given in kilograms of mass per unit of swept area. Lightweight small turbines are typically less than 10 kg/m², whereas more rugged machines range from 10 to 20 kg/m². For a sense of scale, Southwest Windpower's Air series of micro turbines is about 6 kg/m² and has the lightest weight of the micro turbines on the international market.

Results similar to Summerville's on Beech Mountain were also found at the Wulf Test Field in California's Tehachapi Pass. There, a mini wind turbine—a Whisper H40—operated for four years, reliably charging batteries with no problems, until it failed.

For micro and mini wind turbines, it may be reasonable to expect a lifetime of half a dozen to a dozen years.

One of the problems limiting the effectiveness of small wind turbines in particular is that they are, well, small, and consequently don't produce a lot of energy. All wind turbines eventually need repair or service, some more than others. As Scoraig Wind Electric's Hugh Piggott points out, what's the point of repairing a small wind turbine if it doesn't generate enough electricity to justify the cost? "Where is the motivation to get it fixed?" asks Piggott. This is a particular concern of Piggott's regarding the small, rooftop-mounted wind turbines that have appeared in Great Britain.

When there's no good reason to fix a wind

turbine, it invariably stands idle. While Europeans might be inclined to remove the turbine out of public embarrassment, North Americans are more likely to leave the turbine standing derelict than to invest in the effort to remove it. In the United States there are derelict wind turbines from the early 1980s still standing idle. They have yet to be removed.

Longevity of Household-Size Wind Turbines

The Bergey Excel is a rugged household-size wind turbine that has been on the market since the early 1980s. Some of these units have been in near-continuous operation since then. West Texas A&M's Alternative Energy Institute has operated a Bergey Excel in the Texas Panhandle for more than 15 years, with only one major repair to the turbine itself and three repairs to its inverter. AEI's Bergey has consistently generated about 9,000 kWh year in, year out at the site near the campus in Canyon, Texas. The turbine delivers about 230 kWh/m² of rotor swept area per year at this site. There is no long-term data on the average wind speed at AEI's test field, but the Texas Panhandle has been long noted for its good winds.

With a little bit of luck, careful monitoring, and maintenance when needed, the more rugged of the household-size turbines—like the Bergey Excel—should last 15 to 20 years.

Longevity of Small Commercial-Size Turbines

For commercial wind turbines, the rule of thumb has been a 20-year life span, and many machines have demonstrated that this is a realistic expectation. BTM Consult reports that many early turbines in Denmark are being replaced before they reach the end of their useful lives. As the technology rapidly advances, explains BTM Consult, a turbine's economic life becomes shorter than its technical life. Consequently, many used turbines from California and Denmark have found new homes, showing that these early machines still have many more years to spin wind into electricity.

Even some of the old turbines from the 1970s and early 1980s continue to spin out the kilowatt-hours. For example, the US Department of Agriculture's experiment station near Bushland, Texas, operates an early version of the Enertech E-44, a product no longer manufactured (see figure 8-1, Enertech E-44 at Bushland, Texas).

Thankfully, the USDA has kept careful records chronicling the turbine's performance for more than two decades. Overall, the turbine has logged more than 100,000 hours of operation, and, on average, has been available for operation 88 percent of the time. The USDA's E-44—a moniker designating the rotor diameter in feet—is a 13.4-meter (44 ft) diameter small commercial-size turbine operated in three different configurations: first 20 kW, then 40 kW, and finally 60 kW. About every five years, the USDA takes the turbine down and gives it an overhaul.

This particular design was notorious for being sensitive to blade soiling, which would substantially reduce its performance. Nevertheless, the USDA's long-term record keeping shows that this early turbine delivers about 500 kWh/m² per year at the Texas Panhandle site where the average wind speed for the period was 5.7 m/s (~13 mph). Importantly, minimum production was as little as 300 kWh/m² per year, while during a year of peak performance the E-44 delivered as much as 650 kWh/m² (see table 8-1, Bushland Enertech E-44).

Figure 8-1. Enertech E44 at Bushland, Texas.

Similar-size turbines in California's windy passes and on the coast of Denmark have delivered equivalent performance over a comparable period of time. Some of these early turbines have been in near-continuous operation for a quarter century—and they're still in operation.

Longevity of Large Turbines

Large wind turbines are designed to last 20 to 25 years, though no multi-megawatt turbines have been around long enough to prove that. The oldest multi-megawatt turbines are still in operation after a decade of service in Germany, and if the longevity of early Danish turbines is any guide, large turbines will last as long as well.

Of course, large wind turbines require regular professional care and service. Major components, such as gearboxes, generators, and blades, will have to be replaced or repaired periodically. But this is true of all large machinery—and these machines are indeed large.

Manufacturers of gearless or direct-drive turbines emphasize that their designs eliminate one of the more troublesome components of large turbines, the transmission. It is noteworthy that Bear Mountain Wind chose Enercon's direct-drive turbines for its large project near Dawson Creek in British Columbia. Enercon is Germany's largest wind turbine manufacturer. The direct-drive machine, considered the Mercedes-Benz of wind turbines, is frequently the choice of German farmers and *Bürgerbeteiligungen* (citizen-owned projects).

More than one-third of Enercon's sales to the German market are to citizen-owned wind projects. They sell another 14 percent to farmers and individual owners. Thus, more than half of

Table 8-1: Bushland Enertech E44				
	Avg. Wind Speed m/s	AEO kWh/yr	Annual Specific Yield kWh/m²/yr	% Efficiency
Maximum	5.6	91,732	649	36%
Average	5.7	71,232	504	27%
Minimum	5.1	43,750	310	23%
13.4 m diameter; 141 m², 20 years of operation				

Enercon's sales in Germany are to small, locally owned projects. Observers have suggested that local or "community" owners want wind turbines that last for the long term without the need for expensive gearbox replacements, and are thus willing to pay a premium price for longevity. Corporate owners, they argue, have a shorter time horizon—especially in the United States, where tax subsidies are limited to 10 years—and opt for turbines with lower upfront costs, which may be more problematic over the long term.

Economics and Profitability

The cost of generating electricity from wind energy is primarily a function of the installed cost, the amount of electricity generated, and the cost of operating and maintaining the wind turbine during its lifetime.

Lower the installed cost and increase the amount of electricity produced, or lower the long-term running costs, and the cost of wind energy decreases. In other words, if the payment for the wind-generated electricity—the tariff—is fixed, profitability increases if you lower installed costs or operating costs or if you increase productivity. The converse is

Wind Turbine Market Survey

The best overall survey of the wind turbine market is published by the German wind turbine owners' association (Bundesverband Windenergie, BWE). It's intended for professionals and sophisticated buyers. In Germany "sophisticated buyers" include farmers, cooperatives, and small consulting companies that specialize in locally owned wind projects.

BWE's survey includes all the essential technical data on turbine types in one handy place, including field-tested power curves, noise measurements, electrical quality, and more. Surprisingly for North Americans, the publication also includes prices and provides comparisons based on cost per square meter of rotor swept area of popular models. The German market is highly competitive, and the data in BWE's market survey is a revealing glimpse of what commercial-scale wind turbines cost in the real world, not the costs appearing in press releases.

The survey is in German, but the technical data uses international units so it's understandable to those familiar with the technology who don't speak German. For more information, visit BWE's Web site at www.wind-energy-market.com.

tors were reporting price increases of 20 to 30 percent. For large turbines in small numbers, relative costs now range from \$2,000/kW to \$2,500/kW, about \$700/m^2 to \$1,000/m^2 for a 2 MW turbine.

The spread among small wind turbines is even greater: from 1,500/m^2 to 2,500/m^2 (see table 8-2, Representative Relative Installed Wind Turbine Costs).

As explained in chapter 3, Estimating Performance, the amount of electricity produced depends upon the wind resource and the turbine's performance. Yields can range from a low of 200 kWh/m^2 to 300 kWh/m^2, for a poorly performing small turbine, to a high of 1,200 kWh/m^2 to 1,400 kWh/m^2 for a large turbine at a world-class windy site in New Zealand.

The annual cost to operate, maintain, insure, and repair large wind turbines in Germany ranges from 4.8 percent of the installed cost per year, during the first 10 years of a 20-year lifetime, to 6.8 percent during the turbine's second decade. These figures are much higher than those bandied about in the press for large wind turbines operating in North America. Unfortunately, the market in the United States and Canada is far less transparent than in Germany and Denmark, and until the market here becomes more transparent it's better to use public data from Europe than press releases from North America.

Large Wind Turbine Sample Economics

In 2007, the Ontario Sustainable Energy Association (OSEA) calculated the price per kilowatt-hour of wind generation that would be needed for a wind turbine to be profitable at a

true, too. Profitability improves if the payment, the tariff, for the wind-generated electricity increases.

In the mid-2000s the relative cost of large turbines, which had been steadily declining for two decades, rose dramatically. German opera-

Table 8-2: Representative Relative Installed Wind Turbine Costs		
	Relative Installed Cost*	
Size	Low $/m^2$	High $/m^2$
Micro	1,500	2,500
Mini	1,250	2,500
Household	1,250	2,250
Small Commercial	800	1,250
Large**	700	1,000

*Gross approximation only.
**Several turbines in small clusters.

Table 8-3: OSEA Estimate of the Tariff Necessary for the Profitable Operation of a Large Wind Turbine			
Average Annual Specific Yield	Average Annual Wind Speed at Hub Height		Tariff Needed
kWh/m²/yr	~m/s	~mph	CAD/kWh
550	5.2	12	~0.15
750	5.9	13	~0.12
1,000	6.7	15	~0.10

moderate site in the interior of the province.[9] It used an estimated installed cost of $700/m² for a small cluster of large turbines that would be cooperatively owned by farmers and the nearby community. This may be a somewhat optimistic estimate at press time in 2009, as costs have continued a steep climb.

At an interior site, OSEA estimated an annual yield of only 550 kWh/m² per year. This is roughly equivalent to an average annual wind speed of 5.2 m/s or about 12 mph at hub height. The association assumed that annual reoccurring costs were 4 percent of the installed cost per year for the life of the 20-year project.

The interior site became OSEA's base case. It also estimated the tariff, or the rate paid per kilowatt-hour, necessary for profitable operation at windy sites. OSEA estimated that a tariff of nearly $0.15/kWh would be necessary in order to operate a small group of turbines in Ontario in 2007 (see table 8-3, OSEA Estimate of the Tariff Necessary for the Profitable Operation of a Large Wind Turbine).

Since large wind turbines have become international commodities, the costs are comparable from one country to the next. It's not surprising, then, that German utilities, under the country's system of renewable energy tariffs, will begin paying a base tariff for interior sites of nearly $0.15/kWh for wind-generated electricity in 2009. The average tariff for windier sites under the new payment schedule will be less than this, but the base tariff indicates that OSEA's calculations were sufficiently robust to reflect real-world costs.[10]

Household-Size Turbine Sample Economics
Let's take a closer look at the numbers for a widely advertised household-size turbine, Southwest Windpower's Skystream. Here are the key assumptions.

- Average wind speed: 5.5 m/s (12.3 mph).
- AEO/AEP: 4,800 kWh/yr.
- Total installed cost: $13,500.
- No subsidies or tax credits are used.
- Total annual reoccurring costs: 4 percent of installed cost per year.

9. See www.wind-works.org/PricingWorksheets/ARTsTariffsPricingWorksheets.html.
10. In Germany, beginning in 2009 all wind turbines will be paid $0.15 CAD/kWh for the first five years. In years 6 through 20, the tariff will fall to $0.08 CAD/kWh at the windiest sites.

There are several other assumptions used in the economic calculations, but they don't affect the results materially.[11]

The average wind speed in this example is better than most will encounter, producing a yield of about 450 kWh/m²/yr. The standard Skystream is installed on a tower arguably too short for the turbine, and especially for Maine, where this price was quoted. If the Skystream was installed on a taller tower, costs would be higher. At this price, the relative cost is $1,250/m². This places the Skystream at the low end of the range of relative costs for wind turbines in this size class (see table 8-2).

Similarly, we're using the manufacturer's estimate of performance. Experience has shown that estimates from small wind turbine manufacturers are often optimistic.

In this example, no subsidies or tax credits are included. At the time *Wind Energy Basics* went to press, there were no subsidies for small wind turbines in Canada, but new tax credits for small wind turbines were enacted in the United States. For those with a tax liability, the US tax credits effectively lower the installed cost of the turbine.

Annual reoccurring costs are expenses for insurance, repairs, and so on. Little is known and even less is published about these costs for small wind turbines. The 4 percent figure is based on the experience of large wind turbines in Germany and is commonly used in public discussions of how high payments for wind energy should be in order to earn a profit. Though the Skystream has been designed for ease of operation and maintenance, there's no long-term operating experience on this turbine.

It would be unwise to assume that its annual costs will be markedly less than for commercial-scale turbines. For the Skystream, the annual reoccurring costs in this example are estimated to be $540 per year. These costs could be lower, but they could also be higher.

What's the bottom line? Using standard economic calculations, owners of this turbine under these conditions could pay for the turbine within the first 10 years of the hoped-for 20-year life of the turbine if they were paid $0.35 to $0.40 per kilowatt-hour for their electricity.

In the United States, the Skystream would qualify for generous federal tax credits if installed at a residence. The Skystream could qualify for tax credits representing as much as 30 percent of its installed cost. This effectively reduces the up front cost of the turbine, making an installation more economically attractive. With the federal tax credit, a homeowner using a Skystream would need to be paid $0.25–$0.28 per kilowatt-hour, a savings of $0.10 per kilowatt-hour over that without the tax credit.

The Special Case of Small Wind Turbines

As seen in the example of the Skystream above, small wind turbines are a special case. They have not progressed as rapidly or as far as large, commercial-scale turbines. Small turbines still offer promise, but a lot of work remains before they become as cost-effective as large turbines. Still, as noted elsewhere, they make sense in specific applications—most notably in battery-charging systems, where they are essential.

11. See www.wind-works.org/SmallTurbines/SmallWindTurbineCostCalculations.html.

US Federal Tax Credits

In October, 2008, the US Congress extended the Renewable Electricity Production Tax Credit (PTC) for commercial wind energy through the end of 2009. And, for the first time in more than two decades, enacted a Residential Renewable Energy Tax Credit that includes small wind turbines. Subsequently, President Barack Obama signed into law an economic stimulus package in February, 2009, that substantially expanded the tax credits and extends the PTC for two more years.

While some details of the tax credits are known, many details remain murky as *Wind Energy Basics* goes to press. For up-to-date information on federal and state subsidies, consult the Database of State Incentives for Renewables and Efficiency (DSIRE) at www.dsireusa.org. For details on the federal tax credits click on the "Federal Incentives" icon.

Most significantly, the stimulus package permits the conversion of some of the tax credits to outright capital grants. This form of subsidy hasn't been seen at the federal level since 1978. Unfortunately, the tax credits, as a form of capital grants on installed cost, are ill-conceived and poor public policy. Apparently Congress learned nothing from similar policies during President Jimmy Carter's administration that led to a reputation for inflated costs and widespread abuses, and that have taken the wind industry decades to overcome.

Small or Residential Wind

For wind turbines installed at a home, whether the principal residence or not, the tax credit was 30 percent of the total installed cost, not including specific state subsidies. The total amount of the credit was limited to the lesser of $4,000 or $500 per 500 watts of the manufacturer's rated power.

Southwest Windpower's Skystream, for example, had been conservatively rated at 1.8 kW. With this rating an installation of the Skystream would have qualified for a maximum tax credit of $1,800. Since the tax credits were instituted in 2008, Southwest Windpower has "rerated" the Skystream to 2.4 kW, an increase in the turbine's rating of 33 percent.

However, the economic stimulus package lifted the previous limitations on the maximum amount of the tax credit. The new law provides a 30 percent tax credit for the purchase and installation of small wind turbines with "rated" capacities of 100 kW or less.

A Skystream costing $13,500 to install would qualify for a tax credit of $4,050. This would bring the after tax cost to about $9,500.

Large Wind

The PTC for commercial sale of wind-generated electricity to the grid is $0.02 pour/kWh for a period of ten years from the date when the wind turbine is placed in service. Thus, if a wind turbine of any size is installed in 2009 and delivers electricity to the grid, it will qualify for $0.02 for each kilowatt-hour produced until the same date in 2019. The PTC is available only for sales to the grid and not to those who net meter. The economic stimulus package allows conversion of the PTC to actual cash grants for those who can't use the tax credits. The cash grants can be up to 30 percent of the project's installed costs. It's likely that most will convert the PTC to cash grants.

There are a host of issues confronting small turbines. Around the turn of the century, researchers at Spain's national energy center, CIEMAT (Centro de Investigaciones Energeticas, Mediombientales y Tecnologicas), took the pulse of the small turbine industry. They were mostly interested in battery-charging turbines, but their findings are relevant to those connected to the grid as well.

CIEMAT's Ignacio Cruz examined the status of small turbine technology, relying in part on an extensive survey of trade literature and in part on measurements at Spain's Soria test field in the highlands northeast of Madrid.

Cruz, the manager of the test field, reached conclusions mirroring those found at small wind turbine test sites in North America. He concluded that small wind turbines were developed principally by "handicraft," and that the "maturity" of the technology was far below that of large, commercial-scale wind turbines.

Small Turbines Cost More

Cruz found that the average cost of small wind turbines for stand-alone, battery-charging applications was also greater than that for grid-connected turbines. At the time costs ranged from €3,500 to €10,000 per kilowatt ($5,000–15,000/kW) for stand-alone installations, in contrast with the €1,000 to €1,350 per kilowatt ($1,500–2,000/kW) for large wind turbines. The specific cost for small wind turbines ranged from €300 to €1,000 per square meter ($500–1,500/m^2), while the specific cost for commercial turbines was about €300 to €500 per square meter ($500–700/m^2).

Small Turbines Are Less Efficient

As explained elsewhere, small turbines typically are less productive than large turbines. Cruz found that this was true as well. Small wind turbines convert less of the energy in the wind to electricity than today's commercial-scale wind turbines. He observed that the maximum conversion efficiency of small turbines from 100 W to 5 kW varied from 0.22 to 0.31. In contrast, large wind turbines reach conversion efficiencies nearing 0.45.

The difference between the two technologies may be even greater. Cruz based his calculations on a survey of power curves found in trade literature. While the power curves for many large wind turbines have been verified by international testing laboratories, very few power curves for small wind turbines have been independently verified. Power curves for many small turbines are inflated.

Measurements made at the Wulf Test Field in California's Tehachapi Pass found some power curves off by 15 to 20 percent at 28 mph (12.5 m/s). For one wind turbine, the power curve was off by more than 30 percent at 15 mph (6.7 m/s). The peak efficiencies for the five turbines tested varied from 0.20 to 0.26.

Room for Improvement

CIEMAT's Cruz concluded that the market for small turbines looks promising, but the technology needs further development. He suggested improvements in reliability, and in passive regulation systems. He also found that "there is a lack of norms, standards, and guidelines" for battery-charging wind turbines, which may retard development.

Small Wind Turbine Test Sites

While there are literally tens of thousands of small wind turbines in use worldwide, there is very little actual field data publicly available. Most information on the performance of small wind turbines is anecdotal in discussions on Internet news groups, such as the one hosted in English by the American Wind Energy Association.[12] Otherwise, data is either not available or private, with its use restricted to the manufacturers that paid for the tests.

Similarly, though there are several international test sites for commercial wind turbines, there are very few catering to small wind turbines. Of sites that test small wind turbines, even fewer publish their findings. Among the professional facilities that do are Appalachian State University and the National Renewable Energy Laboratory. Two amateur test sites also publish their results. Mike Klemen's Web site chronicles his experience operating small turbines in North Dakota. And data on micro turbines from the Wulf Test Field in California's Tehachapi Pass is available on the author's Web site.[13]

Table 8-sb1: Small Wind Turbine Test Sites		
Professional	URL	Reports Available on Web
Applachian State University	www.wind.appstate.edu	Yes
National Renewable Energy Laboratory	www.nrel.gov/wind/	Some
Alternative Energy Institute	www.windenergy.org	No
Folkecenter for Renewable Energy	www.folkecenter.net	No
Amateur*		
Michael Klemen	www.ndsu.nodak.edu/ndsu/klemen	Yes
Wulf Test Field	www.wind-works.org	Yes
*Amateur in the sense that they are not professional laboratories.		

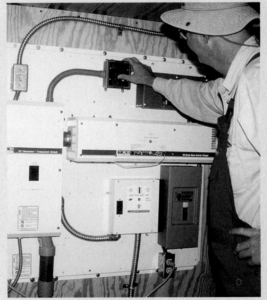

Figure 8-sb1. **Field Testing. Power performance instruments at the Wulf Test Field in the Tehachapi Pass.**

12. See http://tech.groups.yahoo.com/group/awea-wind-home. There's another group for francophones at http://fr.groups.yahoo.com/group/petit-eolien.
13. See http://www.wind-works.org/articles/small_turbines.html.

As a scientist, Cruz limited his observations to the battery-charging turbines he tested. However, he could have made the same observations about most small wind turbines, whether they were intended for charging batteries or feeding electricity into the grid.

Lack of Testing Hinders Progress

Cruz decried the absence of test fields where the "performance and feasibility" of small wind turbines can be measured in accordance with internationally accepted procedures. Such test fields are essential, he noted, for optimizing and improving performance. Standardized testing like that done on large turbines is also critical for providing consumers with the confidence that the turbines will perform as advertised.

Consumer Labeling and Certification

Unlike standardized commercial-scale wind turbines, which have been in existence for many years, standards for small wind turbines continue to evolve. The purpose of both standards and certification is straightforward: to verify manufacturer's claims by reporting the results of standardized testing. Certification is the step manufacturers have to go through to confirm or "certify" that the tests conducted on their wind turbines conform to national or international standards.

Certification is also necessary in certain jurisdictions for a turbine owner to qualify for subsidies and incentive programs. In Britain, certification of small wind turbines will also be a requirement for exemptions from certain planning requirements that apply to almost everything in that crowded isle (see figure 8-2,

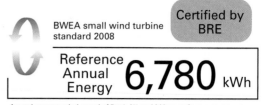

Figure 8-2. BWEA Label. This label indicates compliance with the British Wind Energy Association's small wind turbine standard. (British Wind Energy Association, www.bwea.com)

Proposed label for small wind turbines by the BWEA).

There are three classes of standards: design standards, performance testing standards, and noise measurement standards. Complying with all three and certifying the results is an expensive process that few small wind turbine manufacturers have undertaken.

The design standard is the most confusing to consumers. Compliance with the design standard doesn't mean that the wind turbine won't fail in use, require maintenance, or otherwise experience problems. The design standard simply says that according to existing engineering knowledge, the wind turbine and its components seem right for the application. Compliance with the design standard is not a guarantee. It certainly is not a guarantee that the wind turbine will operate flawlessly for decades, or even that it will deliver the performance promised.

Electrical standards are a subset of design standards. Most wind turbines that are designed for connection to the grid must have the relevant electrical components, such as inverters, tested and certified to a separate electrical standard. Nearly all building codes and most insurers require conformance to an electrical standard. The primary US standard is the National Electrical Code.

Wind turbines using components that are in compliance with an electrical standard, such as those with a UL (Underwriters Laboratories) label, are not necessarily in compliance with any of the other standards. Some manufacturers have tried to falsely link compliance

wind turbines to the same power performance standard as large wind turbines. Similarly, IEC standards hold small wind turbines to the same criteria as large turbines for noise measurements.

In early 2008 the British Wind Energy Association (BWEA) adopted a certification requirement for small wind turbines. Through

---

I've been reporting the "imminent" adoption of testing and reporting standards for small turbines since *Wind Energy: How to Use It* was published in 1983!

---

with electrical standards to a broader approval or endorsement of the wind turbine's basic design. Compliance with electrical standards only suggests that when used in the manner prescribed, the equipment won't cause a fire.

Standards for small wind turbines have been a long time coming. The first draft of the American Wind Energy Association's performance standard appeared in 1979. Nearly three decades after the standard was first proposed, there are still no small wind turbines for sale on the US market that fully comply with its provisions.[14] In early 2008, one manufacturer had certified compliance with the design standards for one wind turbine model, but had yet to certify compliance with both performance testing and noise measurement standards.

Despite decades of efforts, certification standards specifically for small wind turbines have yet to be implemented in North America. As *Wind Energy Basics* went to press, standards of the International Electrotechnical Commission (IEC) applied. These IEC standards hold small

a linkage with Britain's planning system, certification will soon be required in order to gain exemption from the onerous planning process, provided that the turbine is small enough (less than 2 meters in diameter) and quiet enough.

Industry groups say that both standards and measures for certification of small wind turbines, defined as wind turbines that intercept less than 200 m^2 of the wind stream—equivalent to conventional wind turbines 16 meters (52 ft) in diameter—are imminent in the United States. Turbines in this size class are typically rated at less than 75 kW in the Americas.

Buying a Small Wind Turbine for the Home
Many homeowners have no idea how much electricity they consume. Sure, they may know how much money they spend. But that's not the same as knowing how much electricity they consume. Many mistakenly think a micro or mini wind turbine will power their entire home. "Don't expect to spend $500 and become energy-independent," cautions Jason Edworthy,

---

14. See www.wind-works.org/articles/PowerCurves.html for an article describing the decades-long quest for standardized testing and reporting of small wind turbine performance in North America.

What Small Wind Needs to Make Its Mark

The small wind turbine industry needs informed consumers

- Who demand better products and are less susceptible to hype,
- Who know what they are buying,
- Who know what they can expect (we need more tests published),
- Who know what their turbine is producing (metering is a must), and
- Who are willing to pay for quality products.

Many small wind turbines are not yet ready for prime time. To make their mark, small turbines must become as reliable and as productive as large wind turbines.

one of Canada's foremost experts on wind energy. "You have to spend enough money to do it right," he adds—and that's usually a lot more than people expect.

The best place to begin evaluating the cost of a small wind system is to realistically determine what it is you want. Do you want a wind turbine to meet the limited needs of a vacation cabin used only on weekends? Or do you want the wind turbine as a complement to the photovoltaic panels and the stand-alone, battery-charging power system you already have? Sum the anticipated electrical consumption you want the wind turbine to meet, estimate the wind available, and then determine the size of wind turbine you need.

Next, compare the various wind turbines in the size class you need. This is quite subjective,

because you must weigh not only price but also less tangible factors such as quality and reliability. Avoid buying any product on the basis of price alone; instead, look at relative price to determine what is a better buy.

One common—though unreliable—measure is the price of the turbine per kilowatt ($/kW) as used earlier in this chapter. Since there's no standard rating system for wind turbines, products with high power ratings for the same size rotor have the lowest price per kilowatt, but may not produce as much electricity as a turbine with a more realistic power rating. A better measure is the price of the turbine relative to the area swept by its rotor ($/m^2). This measure is helpful because it's not subject to the vagaries of power rating and the size of the wind turbine's generator. That's why table 8-2, Representative Relative Installed Wind Turbine Costs, is presented in terms of dollars per square meter of rotor swept area.

Better yet, estimate the amount of electricity it will produce and compare that with the total installed cost of the wind system. The National Renewable Energy Laboratory now uses what they call a "cost-performance ratio" to evaluate the cost-effectiveness of small wind turbines, says NREL's Jim Green. The ratio is simply the purchase price of the wind turbine divided by its estimated annual energy output. This method eliminates any confusion over various power ratings. To use this method effectively, you need to know your wind resource and how much energy each turbine will produce under those conditions.

As a rule, wind energy becomes more cost-effective as wind turbines increase in diameter. The relative price (price per swept area, or cost-performance ratio) decreases with increasing rotor diameter. Though smaller wind turbines cost less, they are proportionally more expensive.

For smaller, less sophisticated wind turbines, it's also essential to have some means of stopping the turbine when needed. In micro and mini wind turbines, this is often a "brake switch"

As a rule, wind energy becomes more cost-effective as wind turbines increase in diameter.

And always weigh total installed costs, not just the cost of the turbine alone. For micro turbines, a quality tower will cost as much as the turbine—sometimes more. The least costly and most user-friendly tower option is a tilt-up, guyed tubular mast.

To cut costs, some people try to build their own towers. "Don't do it," advises the USDA's Nolan Clark. The tower should be matched to the turbine. Like the rotor, it's a critical component. Several suppliers of small wind turbines offer tower kits that are preferable to building a tower from scratch. These kits require enough home assembly to satisfy the most ardent do-it-yourselfer.

You'll also need all the electronics, if the turbine is not packaged with them. And it's essential to know how the wind turbine is performing, so that when it's not, you can get it fixed. Consequently, all small turbines should have some kind of recording meter so you can track performance over a period of time. Simple meters measuring instantaneous power or current, while helpful, are insufficient for anything more than determining whether the turbine is generating or not. They're better than nothing—but not by much.

that places a large electrical load on the generator, stalling the rotor and dramatically reducing its speed. Unbelievably, many micro and mini wind turbines don't include such essential devices.

Do-It-Yourself Turbines

If you're an inveterate tinkerer and you're determined to build your own wind turbine, contact the Centre for Alternative Technology in Wales or Scoraig Wind Electric in Scotland. Both offer plans for do-it-yourself wind turbines.[15] Scoraig's Hugh Piggott periodically offers workshops in North America in which participants build their own working wind turbines. And yes, these are real wind turbines of a size to be useful, and designed by someone who knows what he's doing. These are not Internet contraptions.

Used Wind Turbines

There's always a limited number of used household-sized turbines on the market.

Check the classified ads in *Home Power* magazine.[16] Unless you've seen the turbine firsthand, and know what you're looking at, it's best to buy only from reputable reconditioners.

---

15. See www.cat.org.uk and www.scoraigwind.co.uk.
16. See www.homepower.com.

Be wary of any used equipment from Californian or Hawaiian wind plants. "They've worked pretty hard," says Wind Turbine Industries' Steve Turek in classic midwestern understatement. "It's like buying a used lawn mower from a golf course." These machines have been beaten to death by those who are pros at squeezing every last cent out of a piece of machinery. Buy a used wind farm turbine only if you've got a strong stomach, and an even stronger bank account.

It's buyer beware in the used wind turbine market. If someone tries to sell you a supposedly "new" turbine at a discount price, think twice. There are no "new" Enertechs, for example. The company went bankrupt two decades ago. If it's a dirt-cheap but "new" Jacobs Wind Energy Systems turbine, call Wind Turbine Industries and confirm that it has in fact been newly assembled.[17] (These are not 1930s-era "Jakes," though they use the same name.) Remember, if the deal you're offered seems too good to be true, it probably is.

There are several reputable dealers of reconditioned Danish wind turbines from California wind farms. These are fully reconditioned and include modern electronic control panels. (The original control panels used relays.) However, these are not discount turbines. For example, a reconditioned Vestas V17—a Danish wind turbine with a rotor 17 meters (56 ft) in diameter—installed with tower can cost about $150,000. In constant dollars, that's as much as it cost when it was installed in 1985.[18]

Nevertheless, used Danish machines have years of life left when put into good hands. Several dozen have found new lives in the Midwest and even in California. There are four old Nordtank turbines operating reliably at the sewage treatment plant in Tehachapi, California, after being removed from a nearby wind farm.

Subsidies and Incentives

Subsidies, or the more politically correct "incentives," are payments or grants given to buyers of wind turbines to encourage them to invest. Typically, subsidies represent some portion of the equipment cost. Often these grants come from taxpayers, and are subject to the give-and-take of periodic budgeting by the legislature.

Sometimes the grants are funded from a pool of money collected from a surcharge on sales of electricity. These Public Goods Funds or System Benefit Charges were created in the aftermath of the deregulation craze that swept North America in the 1990s. Proponents reasoned that when electric utilities were "restructured" there were public goods, like research and development, that would no longer be funded as a part of the new, leaner, and meaner utilities that resulted.

Electric utility regulation—and as a result, Public Goods Funds—are administered at the state and provincial level in Canada and the United States. Consequently, subsidies for wind energy are a hodgepodge of programs. On top of that, the federal governments in Canada and

---

17. See www.windturbine.net.
18. See www.halus.com and www.energyms.com.

How to Evaluate New Small Wind Turbine Technology

Jim Green, a small wind turbine specialist at the National Renewable Energy Laboratory in Boulder, Colorado, offers this simple guide for evaluating new technology.

- What is the performance? Does the manufacturer provide a power curve or an estimate of the annual energy output?
- How was this performance measured? Was it measured in the wind, in a wind tunnel, or on a truck?
- Has this performance been independently verified?
- Does it comply with Underwriters Laboratories 1741 for safe interconnection to the utility (in the United States)?
- Does it comply with the International Electrotechnical Commission's design safety standard?
- Who can provide parts and service?
- What is the warranty?
- Where has it been demonstrated?
- What does it cost? Is the price estimated or based on real manufacturing experience?

And here's what you should expect to find. The manufacturer should be able provide estimates of the actual kilowatt-hours that the turbine will generate on a monthly or annual basis. The turbine should have been tested in the real world—that is, in the wind—and the tests should be independently verified. The turbine should comply with all electrical safety requirements and its design should meet IEC design safety standards. There should be a warranty included with the sale of the turbine, and parts should be available. The turbine and its manufacturer should have a track record. There should be multiple units in operation in a variety of locales and applications. And, of course, any estimates of costs should be based on real manufacturing experience, not guesswork.

the US have their own programs. It's enough to make your head swim.

Legions of lawyers in both countries have carved out profitable niches parsing the complex regulations governing subsidies and the even more arcane tax implications of various forms of wind investment. At least in the United States, there's a database (the Database of State Incentives for Renewables and Efficiency) that tracks all these programs and their requirements.[19] There's no equivalent database for Canada.

Federal Programs

Canada's Ecoenergy renewable energy incentive pays 1 cent per kilowatt-hour of wind generation for 10 years. The payment is only for generation and, thus, of no benefit to those who want to net-meter. While it's a modest—

---

19. DSIRE (Database of State Incentives for Renewables and Efficiency) is a comprehensive source of information on state, local, utility, and federal incentives that promote renewable energy and energy efficiency; see www.dsireusa.org.

some would argue, ineffectual—incentive, the Canadian payment is available to all Canadians regardless of their social and tax status. The Canadian incentive is far more egalitarian than the equivalent American program.

The US federal Production Tax Credit (PTC) is a much ballyhooed, on-again, off-again subsidy program funded by Congress from the federal Treasury. The PTC grew out of an older tax subsidy program of the early 1980s. In the earlier program, the subsidy was based on the installed cost or the capital cost of the wind project. This led to widespread abuses that gave renewables a poor reputation for years afterward. Unscrupulous developers would inflate the cost of the wind turbine so they could collect more subsidies.[20]

Subsequently, when subsidies were eventually restored in the late 1990s they were offered only as a production-based incentive—that is, paid only on the amount of electricity produced. While this eliminated fraud and the tendency to inflate costs simply to increase the subsidy payment, Congress remained timid and placed onerous restrictions on who could use the PTC. As a result, only the most profitable companies, hence those with the most taxes to pay, could participate in the program.

Further, Congress was never certain it really wanted the program, or wind energy, at all. The PTC has been extended, and then allowed to expire, several times in the past. This uncertainty has led to a boom-and-bust pattern that has characterized the US wind industry for the past two decades. As one wag put it, "This is no way to run a business."

State and Provincial Programs

Most state and provincial incentive programs are built around capital or upfront subsidy payments. There are, however, few unique programs in North America. Ontario, Wisconsin, and Washington State have programs that pay a specific amount per kilowatt-hour for the electricity generated.

Renewable Energy Payments (Tariffs)

Ontario pays a feed-in tariff of $0.11 CAD/kWh for electricity generated by wind projects less than 10 MW in size. The program is open to all participants—farmers, homeowners, the province's indigenous population, and businesses. The Ontario program remains the most innovative in North America, but implementation has been hamstrung by an overly bureaucratic administration, and the low tariff. As *Wind Energy Basics* went to press, the government of Ontario was proposing a much more aggressive program modeled after successful programs in Europe. When enacted Ontario could again take the lead in North American renewable energy policy.

In Wisconsin, Xcel Energy (formerly Northern States Power) pays a feed-in tariff of $0.066/kWh for wind generation for a period of 10 years. This tariff is limited to wind turbines from 20 kW to 800 kW in size.

Washington State uses a performance-based incentive coupled with net metering. Because of size limits on what can qualify for net metering, the program is limited to turbines of less than 100 kW in size. However, if the wind turbine and all its components are made

---

20. For more on this era see *Wind Energy Comes of Age* by Paul Gipe, John Wiley & Sons, 1995.

in Washington State, the owner qualifies for $0.33/kWh in incentive payments for seven years, and offsets purchases from the utility valued at $0.08/kWh. Without accounting for inflation, the earnings from such a program are roughly equivalent to $0.18/kWh spread out over a 20-year period.

In 2008 California implemented a rudimentary feed-in tariff policy for wind, solar, and biomass projects. The program is limited in several ways. First, projects are limited to only 1.5 MW for the state's two largest utilities, Southern California Edison and Pacific Gas & Electric, to only 1 MW for all others, and are ultimately capped at 230 MW in total. Tariffs paid under the program vary by time of day and range from $0.06/kWh to $0.31/kWh but average only $0.094/kWh for a 15-year contract.

Capital Subsidies (Grants)

Several states offer upfront subsidies or capital grants, sometimes called buy-downs, that seek to lower the initial cost of small wind turbines. Because these are grants rather than production-based incentives, and the money is paid out whether the wind turbine works or not, there is more opportunity for abuse. Consequently, legislators and regulators place numerous requirements on participants to prove that they qualify for the grant. This leads to a much more cumbersome bureaucracy than with simple renewable energy payments, where the onus is on the investor in wind energy to ensure that the turbine works as expected.

Capital grant programs typically limit the size of qualifying wind turbines to mini or household-size turbines, and often limit the total amount of grant available as well. For example, a program might offer a subsidy of 25 percent for an installation of a small wind turbine, up to a total project cost of $10,000. In the Skystream example used previously, the subsidy would be limited to $2,500, reducing the installed cost from $13,500 to $11,000. With the reduced cost, the Skystream could be operated profitably if it could sell its electricity for $0.28 to $0.33/kWh.

In many cases, the wind turbine model chosen must be "certified" in some manner in order for the owner to participate in these programs. The certification is often lax and frequently ineffective, but it does weed out the Internet wonders and those wind turbine models with no track record whatsoever.

The regional government of Rhône-Alpes in the south of France has a requirement that could be useful for such programs in North America. To qualify for the subsidy, the small wind turbine installation must include metering to measure the turbine's performance, as well as wind speed, and the owner must submit the data to the government administrator.

Cooperative Investment

Despite the vastness of North America, not everyone can install a wind turbine—of any size—in their own backyards. The majority of North Americans live in urban agglomerations or tract housing where a personal wind turbine is out of the question. Even in rural areas, farmers and other small business people may find installing their own wind turbine to be too great a personal risk.

Because of their higher population density,

continental Europeans long ago confronted a similar dilemma. Their ingenious solution was to join together and buy one or more large wind turbines as a group and place them to best advantage in or near their community. This simple but revolutionary idea powered first Denmark, then Germany, to world dominance in wind energy development.

larger pool of risk capital than any one single homeowner. The group or partnership could afford to hire the professional engineers and meteorologists necessary to ensure a sound investment in a large wind turbine. These are the professional skills that are prohibitively expensive for a homeowner installing a household-size turbine.

Thus shared investment enables investors to gain the economies of scale that accrue from using a large wind turbine as opposed to a small one, and access to the professional skills needed for a sound, long-term investment.

The advantages of shared investment over individual ownership were manifold, but most importantly it spread the risk to a larger group and minimized the investment any one family needed to make. Many Danish cooperatives, like the Middelgrunden co-op outside Copenhagen, deliberately offered shares for less than the equivalent of $1,000 to attract as many participants as possible. By doing so, they spread not only the risk but also the opportunity to profit from wind energy to the entire community.

An investment group, whether a limited partnership or a cooperative, could also raise a

Thus shared investment enables investors to gain the economies of scale that accrue from using a large wind turbine as opposed to a small one, and access to the professional skills needed for a sound, long-term investment.

An unsung benefit of taking on such a daunting task as installing a wind turbine is the mutual support and encouragement a group of like-minded investors provides. There's a certain amount of institutional inertia or stubbornness necessary to install a wind turbine— of any size. Large corporations are successful at developing wind energy, in part because

Table 8-4: Representative Shared Ownership Wind Projects							
	Region	Country	MW	million kWh/yr	Investors	Equity million	Total Investment million
WindShare	Ontario	Canada	0.75	1.2	425	$ 0.8	$ 0.8
Regiowind Freiamt	Baden-Whrttemberg	Germany	3.6	5.7	142	$ 2.1	$ 6.5
Minwind I & II	Minnesota	USA	3.8	11.1	66	$ 1.1	$ 3.5
Regiowind Freiburg	Baden-Whrttemberg	Germany	10.8	16.8	521	$ 6.7	$ 20.6
Paderborn	North-Rhein-Westfalia	Germany	18.2	31.4	91	$ 6.0	$ 28.5
Middelgrunden*	Zealand	Denmark	20	44	8,500	$ 38.2	$ 38.2
*20 MW of 40 MW project developed by the cooperative, ~$1,000/share.							

Paying for Performance

Tom Starrs, an attorney with an illustrious career in renewable energy that began during California's famed "wind rush," has observed that during the early 1980s most renewable incentives in the United States were based on capacity payments or as a percentage of installed costs. "The consensus among analysts," says Starrs, "is that the use of capacity-based incentives in the US during the 1980s contributed significantly to performance-related problems and in some cases to outright fraud." He goes on, "These problems contributed to the federal government's abandonment of wind energy incentives and most solar energy incentives in 1986."

Unfortunately, Congress disregarded this experience when enacting the new Residential Renewable Energy Tax Credit for household-size installations in 2008. Like the ill-fated programs of the 1980s, the new tax credits are based on the capacity or power rating of the installation.

Starrs's remedy? Paying for performance. Good public policy pays only for performance, he says. It is in the best interests of ratepayers (electricity consumers) as well as renewable energy proponents that premiums or incentives paid for renewable energy result in the actual generation of electricity. This can only be achieved when payments are coupled directly with the number of kilowatt-hours produced. This places the responsibility on the manufacturer of the wind turbine, and on the investor, where it should be, not on the ratepayer or the taxpayer.

they can financially support a staff who can keep plugging away month after month to see a project to fruition. And it takes nearly as much effort to push a project with 10 turbines through to completion as it does a project with 50 or 100.

Shared ownership can take two forms: a cooperative, or a limited liability company. In a cooperative, the members individually provide all the capital, both equity and debt. The members may borrow the money from their bank to make the investment, but their purchase of shares represents their total investment in the project. For example, coop investors in Middelgrunden's 20 MW share of the 40 MW offshore project raised $38 million (see table 8-4, Representative Shared Ownership Wind Projects). Similarly, co-op members in Toronto raised all the capital for their portion of the WindShare turbine on the harbor front.

Most shared-ownership projects in Germany and those in the upper Midwest use limited liability partnerships. In these projects the owners raise the equity portion of the project and finance the remainder of the project with debt, often a traditional loan from a bank. The equity portion is typically 20 to 30 percent of the project's total cost.

Due to the arcane tax laws in the United States, and Congress's restrictions on use of the PTC, communities in the upper Midwest developed complex financial transactions that allowed farmers and others to build projects while using the tax credits. Several Minnesota farmers developed the Minwind projects, for example, by using what has been dubbed the "Minnesota flip." In these transactions, farmers partner with someone who can use the tax credits. The partner then "owns" the project for the first 10 years, the period during which the tax credits are offered. After the tax subsidy has been exhausted, the partner "flips" ownership back to the group of farmers who originated the project.

Thus, while large commercial wind farms owned by some of the world's largest utilities are being built in Minnesota, nearly one-third of the installed wind capacity has been installed by industrious farmers working together to cycle the revenues from the wind blowing over their properties back into their own communities through their local banks and other institutions.

The Challenge

9

North Americans have been dabbling around the edges of energy policy for nearly three decades. Until recently, few have acknowledged the seriousness of the challenges facing the continent. The continent is vulnerable. It's vulnerable to peak oil—when supply can no longer keep up with demand—because so much of our transport is provided by cars and trucks. It's vulnerable to peak natural gas, because so many of us use it to heat our homes. It's vulnerable to climate change, as our summers become hotter and drought repeatedly stalks our arid western states and provinces.

Though no one can say with certainty that the killer heat wave that hit Europe in 2003 was due to climate change, it did fit the pattern that's expected in a hotter world. Whatever the cause, the result was disastrous. More than 50,000 excess deaths were attributed to the heat, half of those in France alone. Nuclear power plants were forced to shut down across the continent, as cooling water exceeded safe operating temperatures.

Meanwhile, the global price of oil started an inexorable climb to heights not seen before. No one knows where it will end. For nearly two decades oil prices were relegated to the business pages of our newspapers—if even that; now movement in the price of oil has become a near-daily front-page event. The price of oil is again the topic of conversation around the watercooler or Starbucks counter.

Natural gas, long the choice for new power plants in North America, has followed its fossil-fuel cousin in attracting worried attention. Even more volatile than oil, the price of natural gas has exceeded prices not projected by "experts" to be reached 20 or 30 years from now. At times the price of natural gas has been so high that the cost of electricity produced by it, the fuel cost alone, has exceeded the cost of electricity from wind energy.

Houston, we have a problem. The public in both Canada and the United States understands this. They're living it daily now. While politicians in both countries have been hamstrung by futile ideological struggles, their constituents have been trying to figure out what to do. Crowds for lectures on renewable energy in the past few years haven't been this large in nearly two decades—since the last energy crisis. And they want action.

North American Consumption

First, the bad news. North Americans consume more electricity per capita, per unit of gross domestic product, per almost any unit of measure, than any other people on the planet.

We're the world's energy hogs. Of course, this isn't new. Environmentalists and energy security analysts have been saying this for years. Environmentalists have been concerned because of the pollutants such as global warming gases that result from squandering electricity. Energy security analysts have become increasingly concerned because our profligate consumption makes us vulnerable to supply disruptions and the volatility in the price of fossil fuels as they become increasingly scarce.

Now the good news. We have lots of room for cutting our consumption of electricity dramatically. Energy-efficiency guru Amory Lovins has been saying for decades that we can cut our consumption in half. Indeed, we can. We have only to look to Europe to see that it has already been done elsewhere.

Lots of books have been written comparing electricity consumption in North America with that of the rest of the world, so there's no need to repeat that work here. Instead, we'll compare how many wind turbines it takes to meet the electrical needs of a North American household and those of a European household. We'll use rotor swept area as our unit of measurement. This is useful for gaining a sense of scale. Then we'll turn our attention to meeting the challenge facing us.

Wind Turbine Rotor Area Needed to Meet Consumption

From a wind energy perspective, it takes much more of the wind stream—for a given wind speed—to meet the energy needs of a North American home than it does, say, for a German home. Table 9-1, Typical Household Electrical Consumption, offers some round numbers to

Table 9-1: Typical Household Electrical Consumption	
	kWh/year/household (approximate)
Texas	12,000
Ontario	10,000
California	6,500
Germany	3,500

keep in perspective: annual electrical consumption per household.

Let's use our example from earlier and assume we'll use a large wind turbine installed at a modestly windy site, with an average wind speed at hub height of 6 m/s (13.4 mph). An 80-meter (260 ft) diameter wind turbine will generate about 4 million kilowatt-hours per year at such a site. We could calculate how many homes such a wind turbine would serve—and this is exactly what most reporters want to know when they're writing a story about wind energy. It should be clear that the answer to this question is that it depends on where the wind turbine is located. The same wind turbine in Germany will provide electricity for considerably more homes than it will in Texas, for a given wind resource.

But let's turn the question around and ask how much swept area of the wind stream—effectively, how much of a wind turbine—we need to meet the demand. This is useful for visualizing how reducing our consumption reduces the number of wind turbines we must build to meet our needs. Fewer wind turbines needed means fewer environmental impacts, less steel and fiberglass required to build the turbines—all in all simply less to do the job. And if the job is smaller—that is, fewer wind turbines—we can do it much more quickly than

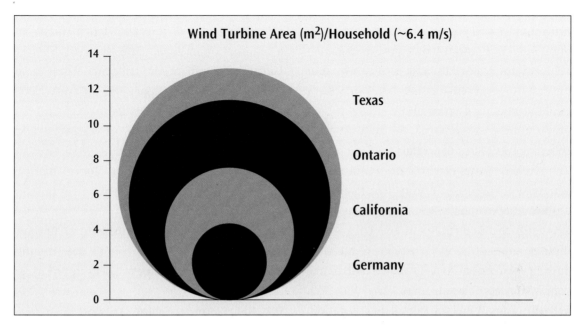

Figure 9-1. Swept Area Needed per Household. The amount of rotor swept area needed to meet average residential electrical consumption.

otherwise (see figure 9-1, Swept area needed per household).

At our hypothetical site, it would take 4.5 m² of the wind stream to meet the needs of the typical German household. To meet the needs of a household in Texas we would need more than 15 m² of the wind stream—3.5 times more area than in Germany—to do the same job.

This should be a sobering thought for those who worry that massive development of wind energy will ravage the continent. If we are going to use wind energy in North America—and we are—and if we're concerned about the impact of wind energy on the landscape, there's no more powerful way to reduce that impact than by reducing our consumption of electricity. Less consumption means fewer wind turbines. It's that simple.

The Challenge

In the summer of 2008, former vice president Al Gore laid before Americans a bold challenge—that the United States meet 100 percent of its electricity from renewable energy and truly clean carbon-free sources within 10 years. He likened it to President John F. Kennedy's challenge to go to the moon, which galvanized the nation at the height of the space race with the Soviet Union.

Despite being immediately dismissed by his critics as a dreamer, Gore's bold vision stirred Canadians and Americans alike. Could it be done? people asked. And if it can be done, could such a national—even continental—endeavor address the dual problems of climate change and our dependence on fossil fuels? Just as importantly, could such a vast undertaking revitalize the industrial heartland of the continent, bringing jobs and economic prosperity back to once proud but now decimated communities?

Clearly, the scale of the task is enormous. But the first question remains, can it be done? Let's examine whether it's possible.

Offsetting Fossil-Fired Generation

The United States consumes ~4,000 TWh per year (4,000 billion kWh/yr) of electricity; Canada, proportionally less. (Canada has about one-tenth the population of the US.) About three-quarters of US electricity is produced by burning fossil fuels. Canada, because of its abundant hydroelectric power, uses "only" 150 TWh per year of fossil fuel for electricity generation. Consequently, we would need to generate about 3,200 TWh per year to offset fossil-fired generation in North America (see table 9-2, North American Wind Energy Challenge: Eliminating Fossil-Fired Generation).

Most wind turbines operating in North America today have been installed on the windiest sites possible. These turbines are highly productive. However, as the industry expands across the continent, it will use increasingly less windy sites. In the example used elsewhere in *Wind Energy Basics,* one 2 MW wind turbine will produce about 4 million kWh per year at a modestly windy site. And this is exactly what we see from large fleets of wind turbines on a regional or national scale, such as those in Germany, Denmark, or California. These large fleets generate ~2 TWh per year for every 1,000 MW of wind capacity installed.

To generate ~2,000 TWh per year, we would need to install a fleet of ~1 million MW. Thus, to offset fossil-fired generation in North America, we would need a fleet representing 1.6 million MW. Using today's 2 MW turbines, that represents some 800,000 wind turbines. Daunting? Yes. But before we move on to determining whether it can be done or not, let's up the ante.

Offsetting Passenger Vehicle Transport

If we're serious about reducing North America's contribution to climate change, and if we're serious about reducing North America's dependence upon oil for transport—with the vulnerability that entails—we need to look at transportation, specifically passenger vehicles. See figures 9-2A and 9-2B.

North Americans drive their cars and light trucks nearly 5 trillion kilometers (~3 trillion miles) per year. Let's say that we were able to convert our light vehicles to electric cars—how much more electricity would that require? Converting our passenger cars to electric vehicles would require an additional 1,200 TWh per year. We'd need another 600,000 MW of wind turbine capacity to meet that need (see table 9-3, North American Wind Energy Challenge: Passenger Vehicle Transport).

Where do we stand now? It looks like we'll need somewhat more than 2 million MW of wind capacity to replace fossil-fired electricity generation and convert passenger vehicle transport to electricity. Of course, we'd need a

Table 9-2: North American Wind Energy Challenge: Eliminating Fossil-Fired Generation		
	North American Fossil-Fired Thermal Generation	Equivalent MW of Wind Turbines @ 6 m/s*
	TWh/year	MW
Canada	150	75,000
USA	3,000	1,500,000
Total	3,150	1,600,000
*@ 2 million kWh/MW/year.		

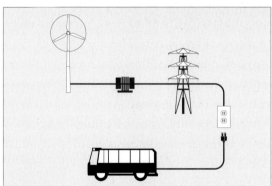

Figures 9-2A and 9-2B. Electric Vehicle Charging. Renewable generation on the grid from wind and solar energy can be used to charge electric vehicles or plug-in hybrids. This role of renewable energy in transportation will become increasingly important as the cost of liquid fuels, such as gasoline and diesel, rapidly rises. Some communities in North America have taken the first, baby steps in that direction. Tehachapi, California, launched a "ride on the wind" program using electricity generated by city-owned wind turbines to charge this electric vehicle.

lot fewer wind turbines if we cut our electricity consumption dramatically and reduced the amount of personal transport by moving toward more public transit.

Manufacturing Capacity

Does North America have the manufacturing capability to build this much wind capacity in any reasonable amount of time? Is it hopeless? No, not at all.

Skilled workers in Canada and the United States churn out 365,000 heavy trucks every year. The steel and other components used in a heavy truck, and the skill necessary to design, manufacture, and assemble it, are roughly comparable to the components and skill necessary to produce a wind turbine. Each heavy truck represents about 500 kW of power capacity, or half a

Table 9-3: North American Wind Energy Challenge: Passenger Vehicle Transport			
	Light Vehicles trillion km/ year	Electric Vehicle Consumption TWh/year*	Equivalent MW of Wind Turbines @ 6 m/s**
Canada	300	75	37,500
USA	4,500	1,125	562,500
Total	4,800	1,200	600,000

*@ 0.25 kWh/km traveled.

**@ 2 million kWh/MW/year and 0.25 kWh/km

MW each. Thus, every year the existing heavy truck industry produces 180,000 MW of power machinery.

What earlier seemed a daunting task now looks far more achievable. If we were to convert our heavy truck industry to building wind turbines in a crash program, we could theoretically build enough wind turbines in less than two decades to offset not only our fossil-fueled power plants but also our dependence upon oil for passenger transport (see table 9-4, North American Heavy Truck Manufacturing as Proxy for Large Wind Turbine Manufacturing).

Wind's Variability

Wind generation is variable. At any single wind turbine, the wind is not always blowing. However, when a continent-wide network of wind turbines is connected, wind can provide

a significant portion of total generation without disrupting our much-appreciated reliability of supply. Some studies have suggested that 50 percent of supply can be provided by wind with modest amounts of backup generation. One recent study by the Department of Energy suggested that the United States could absorb 1 million MW of wind capacity within the existing grid at only modest cost.

Wind energy is the focus of this book, but wind is only one form of renewable energy. To build a truly sustainable supply we will need all forms of renewable energy, not only wind. We can't forget that solar, geothermal, and biomass will play an important role as well. Coupling wind with other renewable resources dampens the cycle of each. Some, such as geothermal and new concentrating solar plants with storage, can be called upon to follow load and pick up the slack for other renewables. New studies have found far more geothermal resources in North America than once imagined. Opening up new possibilities, too, is the rush of plans for large solar plants in the desert Southwest.

Another possibility that has long been talked about is the greater integration of continental transmission networks, to better link east to west and north to south. Such links could enable

Table 9-4: North American Heavy Truck Manufacturing as Proxy for Large Wind Turbine Manufacturing					
	Units/year	Equivalent MW of Wind Turbines/year*	Years to Meet Fleet Required to Offset Fossil-Fired Generation	Years to Meet Fleet Required to Offset Passenger Vehicle Transport	Total Years
Canada	65,000	30,000	2.5	1.3	3.8
USA	300,000	150,000	10	3.8	13.8
Total	365,000	180,000	8.9	3.3	12.2

*@ 0.5 MW/unit.

Canada's existing hydro system to provide a valuable balancing role for a continent-wide network carrying renewable generation to and fro. Add to this the seemingly futuristic possibility of using the batteries in millions of electric vehicles as envisioned here—to provide storage at the site of demand—and there doesn't seem to be any insurmountable obstacle to replacing three-fourths of the continent's electricity supply—existing fossil-fired generation—with renewable energy.[21]

Affordable

Can we afford to develop such a staggering amount of wind energy? Many would argue that we can't afford not to. We are spending enormous sums to import fossil fuels—money that is lost to our economy, money that enriches others, not North Americans. If we spent these same amounts on developing our own indigenous renewable resources, the money would circulate through our economy, enriching our own industries and our own communities.

Can we afford to develop such a staggering amount of wind energy? Many would argue that we can't afford not to.

Land Area Required

There's more than ample land area in North America for such a large number of wind turbines. Open arrays of wind turbines in a wind power plant—for example, wind turbines 8 rotor-diameters apart across the wind by 10 rotor-diameters from one another downwind—can support about 4 MW per square kilometer (10 MW per square mile). To reach 2 million MW would require some 560,000 square kilometers (225,000 square miles), or about 7 percent of the lower 48 states—less than 4 percent if we cut our electricity consumption in half. And of this land, the wind turbines would use only about 5 percent for roads and ancillary facilities; the remainder would remain in existing uses, whether row crops or grazing.

Again, for a sense of scale, let's look at another big recent expenditure and see how it stacks up to re-industrializing our manufacturing industries with massive wind development. US citizens have spent $550 billion in direct costs through mid-2008 on the war in Iraq. That's almost $2,000 for every man, women, and child in the nation. Consider what that means. Every person in the United States would now—not tomorrow—have 1 kW of their very own wind generating capacity. That's like every family having a 2.5 kW wind turbine in their backyard. Imagine if they were paid a modest 10 cents per kilowatt-hour for the generation from such a wind turbine at a modestly windy site. Every family would earn $500 per year for the next 20 years for a total of $10,000 over the life of the wind turbine. Instead of losing $5,000 per family on the tragic Iraq adventure, we could have invested

---

21. For more on the grid integration of wind energy, see www.wind-works.org/articles/GridIntegrationofWindEnergy.html.

Comparison of Different Wind Energy Targets

In addition to the proposal here for the massive development of wind energy in North America, several other prominent groups have proposed their own visions. As explained in the text, the example in *Wind Energy Basics* is for both Canada and the United States and also includes passenger vehicle transportation. The other plans only consider electricity supply and only in individual countries.

Canadian Wind Energy Association (CanWEA)

CanWEA has proposed meeting 20 percent of Canadian electricity supply by 2010. This is a very modest 10,000 MW.

American Wind Energy Association (AWEA)

AWEA's target is the most conservative of the American proposals, no doubt reflecting the views of the electric utilities on its board of directors. AWEA's target is 20 percent of the US electricity supply from wind energy by 2030. This is equivalent to 300,000 MW.

Pickens's Plan

T. Boone Pickens made his fortune in natural gas. He now has set his sights on wind energy. In the run-up to the 2008 US presidential election, the Pickens organization planned to spend $58 million advertising his proposal in print, video, and electronic media. Ads for the "Pickens Plan" were ubiquitous in the fall of 2008. They were everywhere.

Pickens proposed that the United States produce 20 percent of its electricity from wind energy, freeing up natural gas for transportation. This would require about 400,000 MW of wind generating capacity if the productivity of the wind turbines were comparable to those used in the example in the text.

To Summarize:

- CanWEA (Canada): 10,000 MW
- AWEA (US): 300,000 MW
- Pickens Plan (US): 400,000 MW
- *Wind Energy Basics* (US and Canada): 2,200,000 MW
- Half *Wind Energy Basics* (US and Canada): 1,100,000 MW

$5,000 in our own people, and earned a profit of $5,000 in the bargain.

Here's another way to look at it. If we had invested that $550 billion in direct costs, we could have installed 275,000 MW of wind capacity, enough to meet nearly 15 percent of the current US electricity consumption. Or we would have been one-eighth of the way to our goal of 2 million MW of wind capacity in North America.

Another staggering expenditure, once unimaginable, is Congress's $700 billion rescue package for the American financial system. The combined cost of the war in Iraq and the estimated cost of the bailout package could reach $1.25 trillion. If we had invested this amount directly in wind turbines, we could be generating more than 30 percent of our electricity today from wind energy. If we had used the money to offset the costs of a nationwide system of feed-in tariffs, the United States would be producing 40 percent of its electricity from wind energy—today, not tomorrow.

Not only is such an ambitious undertaking possible, it's eminently doable. No, it certainly won't be easy, or cheap, or without controversy, but it can be done. The question becomes, do we want to rise to the challenge and in doing so transform our society?

North Americans have risen to great challenges in the past, and we can do so again. North Americans built great public work projects to pull ourselves out of the Great Depression. The vast hydroelectric projects on the Columbia, the Colorado, the Niagara, and the St. Lawrence are witness to what we can accomplish when we put our minds to it. We rose to the challenge of fighting fascism in World War II despite those who said the Axis couldn't be beat. We belatedly granted civil rights to all our citizens in the 1960s leading eventually to the United States first African-American President, and in the modern era we have pushed cigarette smoking to the fringes of society when critics said it was a waste of time because it couldn't be done.

We did it. We proved the naysayers wrong. And we can do it again. But we certainly can't do it at the plodding pace of the present, with the timid programs currently on the books. We will have to ramp up development dramatically, and we won't be able to do that without the support of the people, the citizens of both nations.

Everyone has a part to play. We will have to pay for it, surely—we always do in the end, anyway. We'll have to grow accustomed to seeing solar panels and wind turbines in new places, maybe in our backyards, maybe down the street, certainly nearby. We will only gain that acceptance—no doubt grudging at times—if we grant everyone the opportunity to participate, to profit from the renewable energy revolution. And there's only one renewable energy policy mechanism that offers opportunity to all: electricity feed laws.

Electricity Feed Laws

Electricity feed laws are the world's most successful policy mechanism for stimulating the rapid development of massive amounts of renewable energy. Feed laws are also the most egalitarian method for determining where, when, and how much renewable generating capacity will be installed. Feed laws enable homeowners, farmers, small businesses, community groups, and the continent's indigenous population to become renewable energy entrepreneurs. All can sell electricity to their utility company for a profit, whether it's homeowners installing solar photovoltaic systems on their rooftops, farmers installing large wind turbines on their land, or cooperatives building small wind farms in their communities. John Geesman, a former commissioner on the California Energy Commission, suggests that it is this feature of feed laws that makes them so powerful: They can democratize the generation of electricity by distributing the opportunity for ownership to all.

Feed laws, like all policy mechanisms for

spurring the renewable generation of electricity, must, at a minimum, include measures for priority access to the grid and payment of a fair price for the electricity produced. These elements are the two essential parts of the development equation. One without the other will not lead to the kind of rapid growth required to meet any but the most unambitious target. Germany's groundbreaking feed law provided both elements: access and price.

technologies—provide a payment for feeding electricity into the grid that is both fair and reasonable, while encouraging robust growth. Balancing these demands is less difficult than it first appears. Geesman, a former regulator himself, points out that utility commissions in both Canada and the United States have been making just such judgment calls for decades.

The objective is to calculate a tariff based on the cost of renewable generation plus a reason-

Feed laws enable homeowners, farmers, small businesses, community groups, and the continent's indigenous population to become renewable energy entrepreneurs.

The idea is not foreign to North America. In 1978 the US Congress passed the Public Utility Regulatory Policies Act that permitted interconnection of renewable energy generators with the grid. Unfortunately, PURPA didn't specify the price, only the means for calculating it. Feed laws, like PURPA, grant priority to the interconnection of renewable sources of electricity with the electric utility network. Unlike PURPA, however, feed laws specify the tariffs, or rates, that renewable generators are paid for their electricity.

Germany's more recent *Erneuerbare Energien Gesetz* (Renewable Energy Sources Act), for example, in its preamble again clearly provides for access to the grid. The law, one of several examples of Advanced Renewable Tariffs, is formally known as the "act on granting priority to renewable energy sources" and goes on to specify in detail the prices that will be paid for different sources of renewable generation.

Successful feed laws—those that produce the rapid growth of a diverse mix of renewable

able profit. This is in fact how we used to regulate the price of electricity. Regulators were charged with ensuring that utility companies made a fair profit while at the same time protecting ratepayers from price gouging because of the utility's monopoly power.

The difference with past practice is that we determine the appropriate price up front, and then make it available to all comers. We say, in effect, *Here's the tariff. If you can build a renewable energy project and make a profit, go right ahead. The sooner you can build it, the better. We need clean, renewable generation and we need it now.*

Advanced Renewable Tariffs differ from simpler feed-in tariffs by differentiating the tariffs according to several factors. Tariffs within each technology can be differentiated by project size and application or, in the case of wind energy, by the productivity of the resource. There can be several different prices for wind energy, several different tariffs for solar, and so on. What makes the tariffs "advanced" is the increased sophistication in fine-tuning them to achieve the desired objectives. Where we want rapid growth, for example, we increase the profitability by raising the tariff. If we want to spur rooftop solar over ground-mounted solar power plants, we pay more for rooftop solar than we do for ground-mounted systems. This is in fact what Germany and France do with their solar tariffs.

Such tariffs are not a subsidy, explains Jérôme Guillet, a French banker who specializes in financing wind power plants. They offer a fair transaction where the public regulatory authority or elected representatives in effect purchase the guarantee of prices that are capped into the future in exchange for somewhat higher prices today. These tariffs, says Guillet, not only become a hedge against both the volatility of fossil fuel prices and the cost to society of this volatility, but also a hedge against the inevitable increase in the cost of fossil fuels. That's the grand bargain.

Because renewable sources of generation are capital-intensive, they require long periods of time to return their investments and earn a profit. Consequently, the prerequisite for a successful renewable energy program, above all else, is the political desire—the political will—for the program to succeed. And for it to succeed there must be the willingness to pay what it costs for renewable energy generation. Where the will exists, there is the stability of public policy that ensures investors, and the banks that loan to them, that there will be a fair opportunity to earn a return on their investment.

Feed laws don't guarantee a profit. They only guarantee that if the project is built, and it generates electricity, the owners will be paid for their electricity at the price—the tariff—that's advertised. The burden remains on the owner to make sure the wind turbine operates as expected. If it doesn't, or if it doesn't produce as much electricity as planned, the owner suffers the loss.

Successful programs must be simple, comprehensible, and transparent. They must provide simplified interconnection requirements with priority access to the grid. They must provide sufficient price per kilowatt-hour to drive rapid development, and they must provide a contract length sufficiently long to reward investment.

Aggressive targets require aggressive programs. Feed laws and the Advanced Renewable Tariffs that make them work are not for the politically faint of heart. We won't get to 100 percent renewables from where we are today with incremental change in response to timid targets. As one political observer noted, "Incremental change will only get you incremental results."

Further, successful programs either have no cap on the program size—or the cap is so high that there is no fear of reaching it in the early years of the program. This discourages gaming and the hoarding of contracts. Of course, a target of 100 percent renewables or a goal of eliminating all fossil-fired generation is the equivalent of no cap on the program—it's the very definition of an aggressive target.

Importantly, renewable tariffs must be sufficiently differentiated to deliver the kind of renewable development from the technology desired in the location desired. There must be tariffs for each technology under a variety of conditions, such as a tariff for small wind turbines as well as large ones. And the tariffs must be high enough to spur development.

Small Wind Tariff

As an example of how aggressive programs must become, consider the special case of small wind. As we've seen, small wind turbines must be paid a high price for their generation in order to have any possibility of earning a profit. The tariff needed is far higher than anyone in North America has considered before. It's not as high as what is needed for solar photovoltaics—the most expensive of the new renewable technologies available today—but a small wind tariff must be far higher than that necessary for its bigger brother, large wind.

France, Germany, and Spain specify tariffs for a host of technologies, but not for small wind turbines. Switzerland introduced its system of Advanced Renewable Tariffs in 2008. This was the first program to offer a specific tariff for small wind: $0.20/kWh. In 2007 several midwestern states introduced bills into their assemblies calling for Advanced Renewable Tariffs; these bills included a tariff for small wind turbines of $0.25/kWh. (As *Wind Energy Basics* went to press in 2009, none had yet passed.)

Higher tariffs may actually be needed. There's surprisingly little real-world performance data on small wind turbines. The Swiss tariff or the proposed Midwest tariffs may be a good place to start, until regulators can make a more informed decision.

Tariffs for Distributed Wind

Fortunately, we know a lot more about the costs and productivity of large wind turbines than those of small turbines. The problem with the existing tariffs in Ontario and California is that they're designed so that one size fits all

conditions. That will never work to spur broad geographic development.

If there's only one price for wind energy and that tariff is low, as in Ontario, then wind development is limited to only certain areas. Commercial developers can move their projects to wherever they like. They will concentrate only on the windiest sites to maximize their profits. Farmers and other landowners, in contrast, are landlocked. Unlike a commercial developer, they can't move to a windy location to install their own turbines. A one-price-for-all policy favors some, but denies opportunity to everyone else. It can also allow some developers to earn excessive profits, above and beyond those needed to encourage profitable development.

Germany and France wanted to avoid concentrating wind development on scenic coastlines or mountain ridges. They wanted to avoid the kind of wind development seen in California's windy passes. Instead, they wanted to distribute wind development across the landscape, to gain more of the benefits of this renewable energy technology by moving the turbines closer to the load—the people. As a result, they pioneered renewable tariffs differentiated by wind resource intensity. They have been followed recently by Switzerland. In all three countries, the tariffs for wind energy vary by the productivity of the wind turbine. The turbine's productivity, or yield, is a surrogate for the wind resource.

The objective was twofold: to lessen development pressure on the windiest sites by enabling development in other, less windy, sites; and to provide siting flexibility. The programs in Germany and France have been successful in spreading development across the landscape of each country. (Switzerland's program is too new to see any results as yet.) While development still favors the windier regions, development is not solely concentrated in the windiest areas. As a result of the German policy, nearly 60 percent of German wind development is now in the interior of the country and has moved away from the coastline. As in Denmark, wind turbines can now be seen in almost every part of Germany.

Germany and France each use a different mechanism for determining site productivity. However, both use a trial period after which the productivity is calculated and a subsequent tariff is determined. Thus, the maximum tariff is fixed, in order to provide a targeted profitability at the targeted sites, but the final tariff paid for more productive—windier—sites declines on a sliding scale as a function of productivity.

In table 8-3, OSEA Estimate of the Tariff Necessary for the Profitable Operation of a Large Wind Turbine, the wind tariffs calculated by the Ontario Sustainable Energy Association embody this principle. OSEA recommended that for the first five years all wind turbines should be paid a base tariff of $0.15 CAD/kWh, after which the specific yield of the wind turbine or group of turbines would be calculated. The tariff for the time remaining in the contract, years 6 through 20, would then be found from a formula that includes a measure for the profitability. In this case, a windy site would receive $0.10 CAD/kWh—or less. Wind turbines at low-wind sites would continue to receive the base rate of $0.15 CAD/kWh. Turbines at moderately windy sites would be paid a tariff less than $0.15C CAD/kWh but more than $0.10 CAD/kWh, on a sliding scale.

This principle must be part of any aggressive renewable energy policy. If we are to see wind development from the windy Great Plains to modestly windy sites around the Midwest and throughout Ontario, we'll need differentiated wind tariffs to make it possible.

Wind tariffs differentiated by productivity can

- Increase distributed generation,
- Distribute wind development across a geographic area,
- Reduce (but not eliminate) development pressure on the windiest sites,
- Reduce (but not eliminate) social friction by spreading development among many sites,
- Increase program flexibility by lessening pressure to get prices exactly right the first time,
- Reduce wind and technology development risk by determining the final tariff after five years of operation,
- Spread opportunity to all, not just to those fortunate enough to live in the windiest locales, and
- Enable fair profits at medium wind sites while limiting excessive profits at windy sites.

Tariffs differentiated by site productivity are a powerful tool for encouraging wind development where it is needed most, near the load—that is, urban areas. And the principle is not limited to wind energy; it could be applied to solar photovoltaics as well. Clearly, solar systems installed in the blistering sun of the Southwest will need lower renewable energy payments than a rooftop solar system on a home in Detroit.

Who pays for these tariffs? We do, of course. The cost of Advanced Renewable Tariffs is spread across all of a utility's ratepayers as a slight increase in the cost per kilowatt-hour of electricity. It's a far more equitable and stable strategy than billing taxpayers, because those who use more electricity pay more for renewable generation. Besides, we are changing the way we produce and consume electricity. We pay for electricity from conventional sources. As we begin to eliminate those, we can reasonably be expected to start paying for our renewable sources of generation as well.

Energy for Life: The Pursuit of an Ethical Energy Policy

Let's get this out of the way first. There are no panaceas, no quick fixes for the continent's energy challenge. It's taken us decades to get into this predicament, and it will take us a long time to get out of it—if we have the will to do so. Electricity feed laws are one very important solution, but only one.

The title of this section, Energy for Life, is a play on the words and the work of the 19th-century Danish theologian N. F. S. Grundtvig. A contemporary of Kierkegaard, Grundtvig's powerful influence helped shaped the democratic, pluralistic, and egalitarian Denmark that exists today. For many Danes, their country's success with wind energy grew from seeds he planted a century earlier.

What does Grundtvig have to do with energy? The answer is at the same time simple and complex. Energy policy should be built upon the framework laid down by Grundtvig more than a century ago: All public policy, in fact all endeavors, should be life-affirming. It's a simple idea that's difficult in practice.

By 2030 Denmark expects to produce 50 percent of its electricity from renewable resources. Similarly, Germany plans to generate 25 to 30 percent of its electricity from renewables by 2020 and 45 percent by 2030. These are not idle "aspirational" goals. Danes and Germans appear to have the will to meet these targets.

Why? In part, because these nations believe they will create long-lasting jobs for their citizens. When the rest of the world decides that it wants wind turbines and solar panels, Denmark and Germany will be only too happy to sell them their products. But they also believe it's the right thing to do. It is not just a question of abstract economics. They feel a moral imperative to meet their aggressive renewable targets.

Thirty years ago President Jimmy Carter asked Americans to embark on a venture that was "the moral equivalent of war." He may have picked the wrong metaphor, but he was right about the moral implications of how we use energy, who has access to it, and where we get it.

The environmental community's interest in wind energy grew out of the ethical dilemma posed by, for example, the strip-mining of coal. If we were to condemn strip-mining for what it does to the land and people of Appalachia, then what would we substitute? Wind and solar energy are a relatively more benign choice. *Relative* is the operative word. There's no environmental free lunch.

Fortunately, there are a number of priests,

rabbis, ministers, and imams who call upon us as North Americans to examine the ethics of our consumptive lifestyle, just as clerics in the past have called upon us to examine our behavior in the more traditional realms of religious teaching. For example, should we raise a moral red flag when a suburbanite drives his Lincoln

the environmental movement proffered new reasons for being miserly.

Within the environmental movement itself, the debate rages over the amount of emphasis to place on the word *efficiency* on the one hand, and on the word *conservation* on the other. Efficiency means, as Bucky Fuller would

---

Energy policy should be built upon the framework laid down by Grundtvig more than a century ago: All public policy, in fact all endeavors, should be life-affirming.

Navigator from his "modest" 3,000-square-foot, cathedral-ceilinged house to pick up a loaf of bread? Does paying for the gasoline used so inefficiently absolve consumers of responsibility for the choices they've made?

Ethics, relative to the consumption of natural resources, is in part, but not solely, the difference between responsible and irresponsible use. We need a lamp to light the darkness so we can curl up with a good book. The lamp in a dark, occupied room serves a purpose; it meets a need. An office building with all the lights ablaze long after everyone has gone home doesn't meet a need. It wastes or squanders a valuable resource, one costly to extract, refine, produce, and deliver.

Implicitly then, there are values at work here. Should we gouge out great furrows in the earth of Montana, or chop the top off a mountain in Tennessee, so that a careless or irresponsible person in Chicago can leave a light on in an unoccupied room? That's certainly not the way many North Americans were raised. Many were taught "Waste not, want not" by parents who lived through the Great Depression and World War II. They learned frugality long before

say, doing more with less. Conservation implies doing without or using less. In the American context, conservation has been a hard sell. Those of us on the front lines are often chided for mentioning the word. "Americans don't want to hear it," we're warned.

Yet efficiency alone, while important, won't take us where we want to go. Conservation or—if that word is unpalatable—stewardship, of a bountiful but finite earth, is critical to our collective future. For example, which of the following solutions is superior? Is it better to modestly improve the efficiency of the Lincoln Navigator from 14 mpg to, say, 18 mpg? That 30 percent improvement is easily and immediately obtainable. Or is it better to conserve gasoline by not building, buying, or using the behemoth in the first place, and driving a Toyota Prius instead? The Prius gets nearly 260 percent better mileage than the Navigator, yet often does the same job—it carries the driver and the loaf of bread. Or is it better still to build real cities where you can walk down the street, pick up the loaf of bread yourself, and talk to your neighbors along the way? Now, how do you calculate the improvement in fuel efficiency of walking?

Paying the monetary price for a Lincoln Navigator—the largest nonmilitary, personal transport vehicle in the world—doesn't absolve the driver of responsibility for the effluent it spews out, the carbon dioxide it emits, or the military establishment necessary to protect its fuel supply. These costs are borne by us all, including those of us who drive small cars, and those of us who walk to pick up our loaf of bread.

There's a moral price tag for using American soldiers to protect our ability to drive Lincoln Navigators. Try explaining those policy choices to the grieving parents of a young man or woman killed in Iraq.

In electricity, this conflict between efficiency and conservation has come back to bite efficiency advocates in the *derrière*. California building standards require all new homes to meet certain standards: the amount of insulation, type of windows, and so on. Good, we say, we've cut the household energy consumption significantly. But what has happened? Our industrial home builders began building bigger and bigger houses. You've seen them. They're monster homes, McMansions. Their ads gloat that with all the energy-saving features, "You too can now afford twice the house you could before."

The result is that we're not just running in place, we're losing ground. This is the social price for the disconnect among efficiency, conservation, and values.

Let's revisit our suburbanite. And let's put her in affluent Palo Alto and propose a new power plant there to meet rising demand from ever more, ever bigger, McMansions. "No way," our suburbanite will say. "Put those power plants in Kern County; they like power plants down there (snicker, snicker)."

What's unsaid, but understood, is that Kern County is plebeian to Palo Alto's patrician world. This cultural gap isn't unique to California. It can be found among Ontario's lakeside "cottagers" and their disdainful view of farmers and the rural communities that want wind farms and biogas plants erected in their midst.

Is there something wrong with those who want or at least accept renewable energy in their backyard? Not at all. There should be a renewable power plant in everyone's backyard—or on their roof, or down the street, or nearby.

NIMBY, or Not In My Back Yard, is just another manifestation of trying to pass the social costs of energy choices—as in the Lincoln Navigator or the McMansion—on to other, and often less politically powerful, groups.

President George W. Bush once said, "After September 11, Americans are reassessing what's really important in life." Maybe so. Ever more, ever cheaper energy doesn't seem quite such a national priority as before, especially now. Still, there are calls to drill in the Arctic National Wildlife Refuge as if this is somehow our patriotic duty, illustrating that despite the president's comment, policy remains dominated by those who view energy as a commodity—as so many of Bertrand Russell's pins.

North Americans are victims of a myth, a belief so all-pervasive that few question it, and fewer still realize it is a belief—a veritable secular religion. Deregulation, restructuring, market liberalization—whatever you call it—was a craze, a long-running one that lasted three decades. But unlike the harmless hula-hoop craze of the 1950s, this craze hurt real people and will continue hurting their descendants for years to come. This fad resembled the tulip mania that engulfed the Dutch in the 17th century, bankrupting the country.

Neoliberalism's chief tenet is that all things good come from the market, that there's a price for everything and that everything can be "monetized." This was Enron's mantra. Unfortunately, an unregulated free market is a fool's paradise where the values most important to us—love, family, clean air and water, a secure future—are either undervalued or not valued at all; and where price—but only the apparent price—is raised onto an altar and we are asked to worship it.

No North American politician has been immune to this siren's song, and we have crashed the ship of state on her shoals. Deregulation's failure offers us an unparalleled, if unfortunate, opportunity to rethink how we produce, consume, and—most importantly—value electricity in our society.

Electricity is a means to an end—a tool for meeting the needs of people—and not an end in itself.

Therefore, we need to envision an electricity system that is sustainable, meets the needs of people today as well as those of tomorrow, and is built upon sufficiency for all, equitably distributed. We need to envision a system that enhances the quality of life for all, rich and poor alike. Such a system is built upon services rendered and needs met, not upon a constant and never-ending growth of supply. We must envision an energy system for people, or, as Grundtvig would say, an energy system for life.

If we choose to do so, we can regain control of our energy destiny and construct an electricity supply system that emphasizes using energy responsibly and with respect for our neighbors and for the environment.

One achievable near-term objective fitting this vision is lowering our residential per capita consumption of electricity to that of Europeans, who share our standard of living.

Per capita, Europeans consume a fraction of the electricity of North Americans, while enjoying the same standard of living. By several measures, Europeans enjoy a higher standard of living than many in Canada and the United States, yet use less electricity, less oil, less energy in general. They often can walk to pick up a loaf of bread, for example.

A young California woman said, "What we waste today, we steal from our children." She meant it literally as a future mother, but she could easily have meant it metaphorically: What we squander today by our excessive consumption, we steal from future generations.

We can do as well as Europeans. Many families in Canada and the United States have shown that it can be done without hardship.

What would happen if everyone did this? Well, for starters, we could cut our greenhouse emissions in half. That alone makes it a worthy goal.

Of course we still need new sources of supply, especially to replace the old, dirty power plants that are still operating. And one way to do that is to offer the people of North America the same deal that Germany, Spain, France, and now even Switzerland—yes, even the conservative Swiss—offer their citizens, by instituting an electricity feed law. Such a law would simply state that you can connect your solar or wind power system to the grid, and, more importantly, would spell out the price you would be paid for your electricity—a price high enough to make it profitable for you, your neighbors, and your entire community. This simple idea is why Germany and Spain are today's world leaders in wind and solar energy.

That's expensive, you say. Maybe; maybe not. It depends upon your values. Germans, Danes, and Spaniards think it's the right thing to do. To them, it's worth it.

We can bring back the luster of North America's industrial heartland, if we have the will to do so. We can build a bright solar future for North America and for North America's children and grandchildren.

Let's put a renewable energy plant in everyone's backyard. Let's create an energy system for life. Let's do it.

We Can Do It

If you want to take part in one of the great national undertakings of our time, get involved. Talk to the legislators who represent you at the local, regional, and national levels. Ask them what they're doing to accelerate the massive amounts of renewable energy that are needed. Ask them what they are doing to give everyone an equal opportunity to participate in the renewable energy revolution. Ask them what they're doing to make renewable energy feed-in tariffs available to everyone.

To sign up for Vice President Gore's climate challenge go to www.wecansolveit.org.

To register your support for electricity feed laws, and the renewable energy payments that make them work, go to the Web site for the North American Alliance for Renewable Energy at www.allianceforrenewableenergy.org.

Individually, the challenge seems overwhelming, but together we can do it.

Index

Note: page numbers followed by f refer to Figures; page numbers followed by t refer to Tables